First Step
to
Quantum Computing

A Practical Guide for Beginners

World Scientific Lectures in Emerging Technologies

Series Editor: Avram Bar-Cohen
(University of Maryland, USA)

Published

Vol. 1 *First Step to Quantum Computing: A Practical Guide for Beginners*
by Javad Shabani and Eva Gurra

World Scientific Lectures in Emerging Technologies - Vol 1

First Step
to
Quantum Computing

A Practical Guide for Beginners

Javad Shabani • Eva Gurra

New York University, USA

Illustrator

Eloise Yalovitser

New York University, USA

W♭ World Scientific

NEW JERSEY · LONDON · SINGAPORE · BEIJING · SHANGHAI · HONG KONG · TAIPEI · CHENNAI · TOKYO

Published by

World Scientific Publishing Co. Pte. Ltd.

5 Toh Tuck Link, Singapore 596224

USA office: 27 Warren Street, Suite 401-402, Hackensack, NJ 07601

UK office: 57 Shelton Street, Covent Garden, London WC2H 9HE

Library of Congress Control Number: 2024030258

British Library Cataloguing-in-Publication Data
A catalogue record for this book is available from the British Library.

World Scientific Lectures in Emerging Technologies — Vol. 1
FIRST STEP TO QUANTUM COMPUTING
A Practical Guide for Beginners

ISBN 978-981-12-5319-5 (hardcover)
ISBN 978-981-12-5412-3 (paperback)
ISBN 978-981-12-5320-1 (ebook for institutions)
ISBN 978-981-12-5321-8 (ebook for individuals)

For any available supplementary material, please visit
https://www.worldscientific.com/worldscibooks/10.1142/12751#t=suppl

Desk Editors: Balasubramanian Shanmugam/Steven Patt

Typeset by Stallion Press
Email: enquiries@stallionpress.com

Preface

Quantum information and quantum computing have evolved into a revolutionary topic where general interest and scientific research have intensified in the recent years. A successful experimental realization of a quantum computer will bring a new technological revolution similar to the advent of the transistor and integrated circuits that form the modern digital world. Many tech companies are investing in quantum computing and are motivated by its potential to solve hard problems. Therefore, a skilled workforce is necessary to help develop this technology as it matures. Remarkably, the field of quantum computing is multi-disciplinary, needing expertise from physicists, mathematicians, engineers, computer scientists, chemists, and, economists and business leaders.

The traditional route to entering this field necessitates a rigorous physics education that involves foundational studies in quantum mechanics, atomic and molecular physics, optics, and more. Of course, these topics also rely on a good understanding of classical mechanics and electrodynamics. As a result, the traditional path delayed teaching quantum computing to graduate school. However, it has been recognized that basic knowledge for entry-level quantum computing can be prepared for a wider audience, especially younger students or professionals who have a growing interest in this field.

This book is a compilation of materials from three years of summer Q-camps in the Shabani group at New York University (NYU) sponsored by Army Educational Outreach Programs (AEOP). Co-author Eva Gurra was trained in this program with a focus on shortening the time to cover foundations and making the course accessible to beginners while still teaching advanced topics and valuable skills. Eva's introduction to quantum computing was in her sophomore year of college. When she first entered the program, she had just finished taking introductory physics courses and a modern physics course where she was first exposed to the strangeness of quantum mechanics.

Eva's journey in quantum computing began with difficult readings and advanced mathematics. However, through interactions with experts at NYU and other students in the program, and coding with IBM Qiskit, she began to better understand quantum computing concepts, such as superposition, entanglement, quantum state characterization, and the effects of noise. After the first summer, she was motivated to continue her study of quantum physics and proceeded to take upper-level courses in classical mechanics, electrodynamics, quantum mechanics, statistical mechanics, and solid-state physics. Eva came back each summer until she graduated from the University of Connecticut. She is now pursuing a doctorate program in physics with a focus on quantum computing and quantum hardware at the University of Colorado Boulder.

The concepts and topics presented in this book serve to create the foundation of forefront research. These introductory topics include qubits, superposition, and entanglement, with a variety of exercises for the reader to dig into the details and check their understanding. We hope to motivate and guide beginners through these necessary prerequisites. The book is aimed at early college students and motivated high school students with an interest in the topic of quantum computing, pretty much like how Eva got started.

About the Authors

Javad Shabani is a Professor of Physics and Director of Center for Quantum Information Physics at New York University. He received his Ph.D. from Princeton University and conducted post-doctoral research at Harvard University and University of California, Santa Barbara. His research interests mainly include developing novel quantum hardware using materials innovation with recent research focus on topological superconductivity and developing voltage-controlled superconducting qubits. He is an active member of quantum education and workforce development in New York area. He is the recipient of the US Army and US Air Force young investigator awards.

Eva Gurra is a physics doctoral student at the University of Colorado Boulder and the National Institute of Standards and Technology focusing on superconducting quantum devices and sensors. She received her bachelor's degree in mechanical engineering from the University of Connecticut and master's degree in physics from the University of Colorado Boulder. Her interest in quantum computing and quantum education began during her time as an undergraduate in Javad Shabani's lab at New York University.

Acknowledgements

The writing and development of this book have been supported and influenced by many people who led and participated in the quantum summer camps at New York University since 2017. We thank the US Army AEOP which has continuously funded the program and given opportunities to high schoolers and undergrads to engage and learn about quantum computing and devices. We gratefully acknowledge Krishna Dindial and Siddharth Shastry for proofreading, editing, and contributing ideas to the book. We also thank Seyed Mohammad Farzaneh, Kasra Sardashti, William M. Strickland, Bassel Heiba Elfeky, Neda Lotfizadeh, and Matthieu Dartiailh for their contributions to the book. We especially thank Dr. Sara Gamble for the encouragement to the authors as well as the inspiring discussions and visits over the years.

Contents

Introduction

Information science and technology are concerned with the storage, movement, analysis, and protection of information. Our everyday life relies heavily on devices, from iPhones to cloud servers, that use information. The total amount of information created in the world's electronic devices has surpassed the zettabyte (1 with 21 zeroes after it). Perhaps, we are producing and distributing more knowledge than we can use and process, which has fueled our enthusiasm for gathering and sharing information and answering difficult questions.

There are still important questions that we cannot answer using current information technology. For example, it is a computationally difficult problem to model and target a drug solution to cure a disease (e.g. cancer) because current supercomputers do not have the power to calculate the energy states of a complicated molecule, let alone search through all possible variations to find a solution. Another difficult problem is weather forecasting, where, for example, it is possible to forecast the path of a hurricane accurately, but because weather conditions are constantly varying, it would take a long time to get a solution and we need the answer before the hurricane happens!

Quantum computing is a thriving and broad multidisciplinary field encompassing aspects of physics, engineering, computer science, and mathematics. With "quantum" as part of its name, it is fundamentally rooted in physics and math, namely the mathematical framework, of quantum mechanics and linear algebra.

Shortcomings of Classical Physics

At a fundamental level, physics aims to understand the laws that govern nature. Before becoming a branch of science, physics was a philosophical subject, which emerged due to a movement towards a rational rather than a religious or mythological understanding of nature. Many of these philosophical concepts emerged in ancient Greece, where the "laws of the universe" and atomism were proposed and drafted by Heraclitus, Leucippus, and Democritus around 500 BCE [1]. During the classical period (500–400 BCE), Aristotle tried to explain how motion occurred, and the idea of gravity with the "theory of four elements": earth, water, air, and fire. This was an effort to establish some fundamental principles or a theory to explain everything physical.

As mathematics advanced, nature could be more rigorously explained with physical theories, developed on paper, and tested through experiments. For example, Archimedes (300 BCE) gave the mathematical foundation for hydrostatics, levers, and pulley systems. This provided useful and practical knowledge that could be applied to advance technology by developing complex systems of pulleys to move heavy or large objects with less need for manpower. Other civilizations, such as in Ancient China (400 BCE), contributed some of the first studies of magnetism and developed the magnetic-needle compass used for navigation and establishing a true north [2]. In the Middle East (7–15 centuries AD), Ibn al-Haytham was regarded as the "father of the modern scientific method" because of the attention to reproducible experimental data. During the 16th and 17th centuries, a scientific revolution enabled progress in the fields of mechanics and astronomy, which resulted in *universally* valid characterizations of motion and planets in the solar system. One of the most prominent figures was Galileo Galilei, who is regarded as the "father of modern science" due to his push for mathematical descriptions of physical phenomena.

By the late 1800s, physics had greatly advanced and expanded to sub-categories of its own: mechanics, electromagnetism, thermodynamics, and statistical mechanics. Classical mechanics, also known as Newtonian mechanics, was developed based on the physical principles laid out by Sir Isaac Newton and mathematical methods invented by renowned mathematicians Gottfried Leibniz, Joseph-Louis Lagrange, and Leonhard Euler. To this day, it remains a powerful theory to predict the motion of macroscopic objects including astronomical objects. Electromagnetism describes the interaction between particles with electromagnetic fields. The theory and principles of electromagnetism, whose roots began with physicists Charles-Augustin de Coulomb, André-Marie Ampère, and Michael Faraday, were combined into compact equations by James Clerk Maxwell. Electromagnetism has formed the basis for explaining light, waves, electricity and magnetism. Thermodynamics explains the nature of heat flow, temperature, and work, and how they are related to energy, radiation, and physical properties of matter itself. Statistical mechanics rose out of thermodynamics to explain macroscopic physical properties such as temperature and pressure in terms of microscopic parameters. Large numbers of particles can be described using probability distributions, and macroscopic parameters can be calculated from these probability distributions. Ludwig Boltzmann, James Clerk Maxwell, and Josiah Willard Gibbs were the key founders of the field of statistical mechanics.

Experiments and theoretical work during the late 1800s and early 1900s showed that classical physics, while very powerful, could not explain all aspects of nature. One of the first shortcomings in classical physics was in the prediction of heat capacities for diatomic gases, such as hydrogen, oxygen, and nitrogen.[1] Classical calculations required that any energy added would be *equally* divided among the different forms of motion (translational, rotational, and vibrational),[2] but there was no minimum energy to mark where a particular motion would "kick in" [3]. This was later explained by quantum mechanics, which allowed for a minimum or threshold energy for rotational and vibrational motions.

Other shortcomings of classical physics were in the concepts of space and time which led to Einstein's special theory of relativity

---

[1]These are very common and abundant gases on Earth, so it was critical to have a theory that matches closely with experimental results.

[2]This is known as the Equipartition theorem. Translational motion is when a molecule moves from one place to another along the same axis. Rotational motion occurs when a molecule rotates around an axis. Vibrational motion is when molecules vibrate *in place*.

proposed in 1905 [4]. In essence, the theory states that clocks in different inertial reference frames run at different rates which depends on their relative velocities. This was confirmed in high-energy experiments involving collisions of elementary particles, namely protons with carbon and other nuclei in the atmosphere, to produce other tiny particles, such as pions or muons.[3] Such high-energy particles are constantly bombarding us due to cosmic rays colliding with molecules in the upper atmosphere. When cosmic ray protons collide with air molecules, pions are formed, which decay into muons after a short time. These muons can be detected because they reach the surface of Earth while pions do not. To measure the average lifetime of a pion, particle accelerators are used to accelerate protons so that they travel about 99.997% the speed of light. When they collide with each other, pions are produced.

When pions are produced in the lab, they are observed to have an average lifetime of 26 nanoseconds. In other experiments, where the pion is moving at about 91.6% the speed of light (about 275000 km/s) relative to the laboratory and had an average lifetime was about 64 nanoseconds [5]. The only difference between the two experiments was the relative motion of the observer and the particle[4] and it meant that time is different for observers in motion than for those at rest. This highlighted a breakdown of Newton's laws, where time is considered to be the same for all observers. Similar experiments showed that space was not the same for all observers as well and that it stretched or contracted depending on the frame of the observer. As a result, velocity could no longer obey classical laws either. In classical physics, there was no upper bound on velocity, but these new properties of space and time dictated that there is indeed an upper bound on how fast particles could travel: the speed of light, 2.998×10^8 m/s.

Another striking limitation of classical physics was in understanding atomic structure. Although atomic theory had existed in a philosophical sense since 500 BCE, it was not until the early 1800s that John Dalton presented the first *scientific* atomic theory, proposing that an atom was a hard sphere that could not be subdivided. By

---

[3]A pion is born through the collision of a proton with a carbon nucleus in the upper atmosphere. Pions are short-lived particles and they quickly decay into muons and neutrinos.

[4]In the "lab frame," we are sitting there watching the pion, while in the moving frame, we would move with the pion.

the beginning of the 1900s, JJ Thomson had discovered the electron and believed that atoms could be divided. Several years later, his student, Ernest Rutherford conducted the so-called gold-foil experiment in 1909, where he showed that the atom consisted of a positively charged nucleus that contained most of the mass of the atom with electrons orbiting it [3]. However, these orbits had no particular structure, and based on classical physics, an orbiting electron would constantly change direction and emit light as it continuously lost energy. This meant that after a very short time, the electron would spiral down into the positively charged nucleus, and therefore, atoms could only exist for a very short time on the order of 10^{-12} s. Obviously, this was catastrophically wrong. If that were the case, matter as we know it could not exist!

Development of Quantum Mechanics

Blackbody Radiation

As the shortcomings of classical physics became more and more apparent, physicists were coming up with ways to explain the discrepancies and in the process create a new theory. The word *quantum* was first used by Max Planck in 1900 when he studied blackbody radiation [6].

Theoretically, a *blackbody* is any object that absorbs all radiation falling on it perfectly and emits radiation more than any other object at the same temperature, as depicted in Fig. 1. The radiation it emits is called blackbody radiation which occurs when the body is

Absorbed   **Emitted**

Figure 1.   Depiction of a blackbody. In the first case, we have a small hole in a large box where all radiation that gets inside bounces around and is absorbed. In the second case, we can think of the box as an oven with a tiny hole emitting radiation that only depends on its temperature.

at thermal equilibrium (has a uniform temperature). Thus, it is considered a perfect absorber and an ideal emitter because it does not reflect or transmit any radiation. At room temperature, a blackbody appears to be black because it radiates infrared light which we cannot perceive. As the temperature increases, it appears reddish and eventually becomes blue-white. Experimental results showed that the blackbody spectrum only depended on the temperature of the object and not the material and that more energy was emitted at higher temperatures for all wavelengths. While classical physics could capture these points, it was unable to account for the shape of the blackbody spectrum which increased quickly, peaked at blue wavelengths,[5] and then sharply decreased for visible wavelengths and beyond, as shown in the graph in Fig. 2.

In classical physics, the concept of an energy limit did not exist, so it predicted that there should be no energy limit for radiation at low wavelengths, but this was experimentally proven wrong. Max Planck proposed that the energy for different frequencies of light should be different. Energy comes in a "clump" known as a **quantum** and is transferred in set amounts directly proportional to the frequency of light.

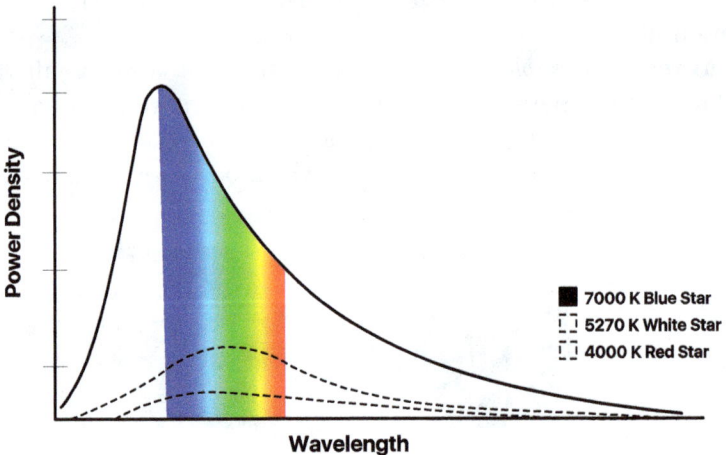

Figure 2. Emission spectrum shape for a blackbody at a temperature of 7000 K.

---

[5]This is why blue stars are hotter than red stars.

Photoelectric Effect

The start of quantum physics is marked by Einstein's formalization of the *photoelectric effect* in 1905 [7]. Experimentally, the photoelectric effect was first discovered by Heinrich Hertz in 1887, who illuminated a metal surface with light and observed electrons being emitted from the surface [3]. To explain this phenomenon, Einstein proposed that light behaved as localized quanta called *photons*. Photons have linear momentum and energy which are characteristic properties of particles. A single electron released from the metal is a result of the electron colliding with a *single* photon which transferred its energy to the electron, as illustrated in Fig. 3. This phenomenon was verified in further experiments with different metals. This work showed that the classical distinction between waves and particles was not correct.

Bohr's Atom

The concept of a bundle or a clump of energy (quantum) helped Niels Bohr rectify the mistake of classical physics in explaining atomic structure and provided the reason for the observed absorption and emission spectra of hydrogen in 1911 [8]. He postulated that electrons

Figure 3. Cartoon view of photoelectric effect. Photons transfer their energy to the electrons in the metal which ejects them from the surface of the material.

revolve in stable orbits around the nucleus without radiating energy. The orbits could be numbered and have limits on the number of electrons they can host. For example, the innermost orbit can host two electrons, the second eight electrons, the third 18 electrons, and so on. The key was that electrons could only jump from one orbit to another by absorbing or emitting light at specific frequencies, as explained by Planck. For a single electron, a *photon* at the right frequency is necessary to jump up an energy level. Likewise, a photon at the same frequency will be emitted if that electron jumps back down. This absorption and emission of photons at specific frequencies is what would determine, for example, the colors that metals appear.

Double-Slit Experiment

It became clear that this new way of thinking provided necessary explanations for observed phenomena, but a theory had yet to be formulated. In 1924, French physicist Louis de Broglie proposed that the concepts of momentum, which corresponds to objects with mass, also corresponded to light or photons which were massless [9]. Likewise, a particle, such as an electron also has some wavelength associated with it. This is all because of the idea of wave–particle duality! In 1927, a diffraction experiment using electrons, conducted by Davisson and Germer at Bell Laboratories, showed that what was classically considered a particle can also exhibit wave-like properties [10]. Typically, when a wave encounters a slit or obstacle, it bends through the slit or around the corners and produces a distinctive diffraction pattern which has peaks of *constructive interference* at specific angles determined by the width of the slit or obstacle and the wavelength.

The experiment involved sending a beam of electrons through a single slit, which was reflected from a crystal of nickel on the opposite side, as portrayed in Fig. 4. The electrons were reflected at some angle relative to the incident beam and some would strike a fixed detector positioned on one side [3]. Based on the size of the slit, it was observed that at some specific angle, the reflection intensity was highest. This was used to calculate the "wavelength" of the electrons and results matched closely with the predicted de Broglie wavelength. Further experiments with different particles showed the same result, verifying the strange *wave–particle duality* concept proposed by Einstein!

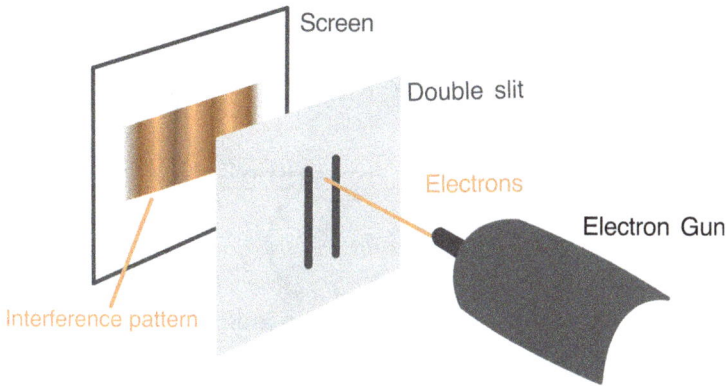

Figure 4. Double-slit experiment with electrons.

The concept of wave–particle duality became one of the hallmark properties of quantum theory. In quantum computing, this concept of interference is important when executing quantum algorithms because you want to boost the amplitude of getting the correct answer.

Stern–Gerlach Experiment

Another significant experiment was the Stern–Gerlach experiment which demonstrated the existence of an interesting property of quantum system: intrinsic angular momentum also known as *spin*.[6] This experiment was developed by Otto Stern in 1921 and conducted successfully by Walter Gerlach in 1922 [11]. The experiment originally involved sending a beam of silver atoms along one direction through a non-uniform magnetic field oriented in a direction perpendicular to the incoming beam and observing the distribution on the screen, as illustrated in Fig. 5. A non-uniform field was necessary so that the silver atom would experience a sideways deflection and hit the screen.

---

[6] Angular momentum is a classical phenomenon and it is the rotational analogue of classical momentum. It can be naturally understood by thinking about a spinning disc.

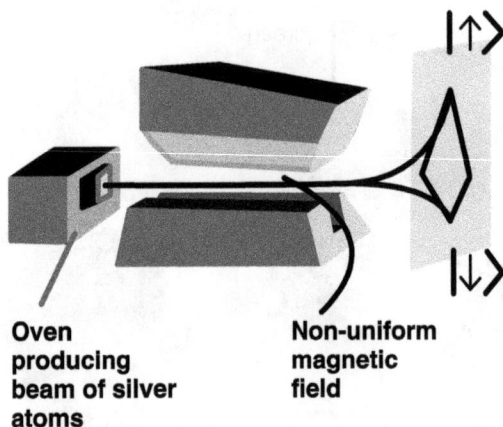

Figure 5.   Stern–Gerlach apparatus.

Classical physics would have predicted a continuous spread (would be represented by a line on the screen) of magnetic moments[7] due to the random thermal effects in the oven that produce silver atoms. However, instead, the atoms would either deflect upwards or downwards, confirming a kind of space quantization known as intrinsic spin.

As we will see later in the book, a single qubit can be described as a superposition of two states, and it turns out that a spin controlled by magnetic fields can be a natural qubit! In fact, spin qubits are currently an active area of research and are promising in terms of scalability.

First Quantum Theory

Experimental developments and theory were working concurrently during these beginnings in hopes of arriving at some concrete rules for how reality works at small scales. Erwin Schrödinger developed his work on the Schrödinger equation in 1926. The equation is known as a linear partial differential equation where the goal is to solve for the wavefunction or the function that fully describes the state of a quantum system. It can be thought of as the analog to

---

[7]The magnetic moment is a vector quantity that represents the strength and orientation of a magnet or any object that produces a magnetic field. Everything from atoms, and cells, to astronomical objects has a magnetic moment.

Newton's second law. Schrödinger's work was fundamental to developing a bonafide theory for quantum mechanics. Many contributors to the theory followed, including Werner Heisenberg, Max Born and Pascual Jordan, who introduced "matrix mechanics". They developed an interpretation for the wavefunction that came to be known as the **Copenhagen interpretation** [12]. The Hamiltonian which describes the total energy of a particular system was represented in matrix form. It was a good formulation to account for the jumps in electron orbits, for example, since matrices describe discrete systems by nature.

The Copenhagen interpretation remains the oldest and most commonly taught interpretation of quantum theory where particles or any material object is ill defined until it is observed or measured. Before measurement, its state is fully described by its wavefunction, and after measurement, the system probabilistically *collapses* to one defined state, known as an *eigenstate*.

Other Quantum Phenomena

Other interesting quantum phenomena include quantum tunneling and entanglement. Tunneling occurs when some particle has a probability of traveling through a barrier without breaking energy conservation. Classically, it makes sense that an object cannot get past some kind of barrier (potential in physics terms) unless it has enough energy to overcome it, but quantum particles have a wave-like nature. For example, sound waves can travel through a wall so that a person can hear it on the other side. Perhaps, it will be more muffled due to the barrier, but some of it still travels through just enough so that we can hear it in the other room. The same thing would happen to a quantum particle in this case which has its own wave-like nature.

Quantum tunneling was experimentally discovered from studies of radioactivity in 1896 by Henry Becquerel. Studying uranium salts, he found that radiation came from the uranium itself without any need for excitation by external energy sources. Along with Marie and Pierre Curie,[8] they studied thorium, polonium, radium, and

---

[8]Unfortunately, the discovery of radioactivity led Bacquerel to suffer serious skin burns from handling radioactive materials and he quickly died in 1908 from acute radiation exposure, a cause that was unidentified at the time. Marie Curie and Pierre Curie followed later in 1934.

other radioactive elements. They discovered and noted that each element had a distinctive decay rate and that decay was probabilistic because of the instability in radioactive nuclei. Particles and photons were being emitted from radioactive nuclei through tunneling via an energy barrier (the nuclei of the radioactive element in this case) that could not have otherwise been crossed if the particle did not have enough energy.

Entanglement occurs when the states are highly correlated, so knowing the state of one entangled particle lets one know about the states of the other particles in the system. When measuring physical properties, such as position, momentum, and spin, of entangled particles, results may be perfectly correlated. For example, if we have two entangled electrons where the total spin of the system is zero and one measurement shows one electron to be spin-up, the other electron is known to be spin down!

Albert Einstein, Boris Podolsky, and Nathan Rosen first discussed this strange prediction of quantum theory in 1935 [13]. In fact, their goal was to show that the newly developed quantum theory involving wavefunctions was incomplete. The thought experiment is known as the EPR paradox. Einstein and Schrödinger were deeply dissatisfied with entanglement due to concerns about information traveling faster than the speed of light travel. This is why Einstein later coined the term "spooky action at a distance" to describe entanglement.

For a while, the paradox remained unsolved until 1964 when John Stewart Bell showed that quantum theory predicted an upper limit on the strength of correlation or level of entanglement, which is known as Bell's inequality [14]. There was no fishy faster than the speed of light communication! By 1972, the first experiments were conducted to test Bell's inequality. The experiment involved measuring the spin of two entangled photons and comparing results with what was predicted by the EPR paradox vs. Bell's inequality.

The mind-boggling aspect of this phenomenon is that the spatial separation of entangled particles is irrelevant because entangled particles form a system. In this system, any one particle cannot be described without considering the others. This work showed that the principle of "local locality" where an object is *only* directly influenced by its immediate surroundings was false. Einstein died before the EPR paradox was resolved although he probably would have disliked this anyway.

Development of Computing

The modern programmable computer started out as an abstract model known as the *Turing machine* [15, 16]. This Turing machine performs computations by reading and writing to an infinite tape based on programmed instructions. The Turing machine consists of the tape that represents memory, a tape head or a pointer to read or write an inspected cell from memory, and a state transition table that gives the instructions, as depicted in Fig. 6.

The machine can only discern the difference between two states, which we now consider the states for a bit 0 or 1, but in an abstract sense, it could be any two states that are different. It operates using a tape head or a pointer which is always on one particular state or memory cell on the tape. The pointer is programmed to move by consulting a transition table which instructs it to either do nothing to the current cell or write a new value to change its state and whether to move left or right. This exactly simulates what we consider modern-day programs which have many lines executed one after the other and depending on the data in memory, the program can produce different results for different data.

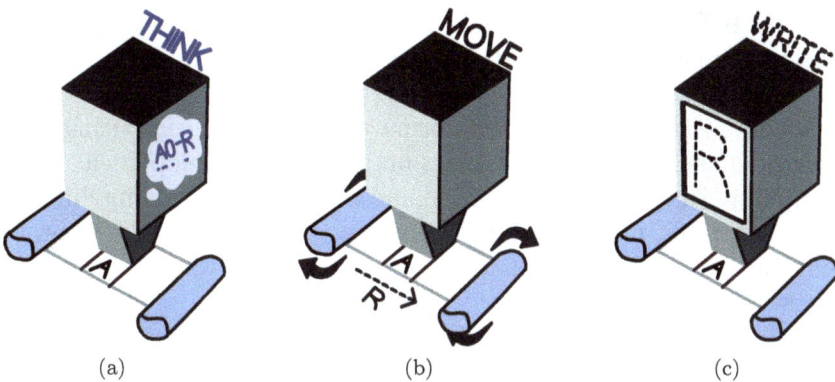

(a) (b) (c)

Figure 6. Representation of Turing machine.

Although this fundamental idea seems so simple now, it was revolutionary at first! It gave rise to the Church–Turing thesis:

> Any algorithmic process can be efficiently simulated using a Turing machine.

The idea is that any machine that can perform algorithms is a machine that is simulated by a Turing machine. Computers from electrical components were soon constructed after Turing laid this theoretical basis. Further, John von Neumann came up with a theoretical model for assembling the necessary components for a computer to act as a Turing machine. By 1947, John Bardeen, Walter Brattain, and Will Shockley developed the transistor [17]. This is the electrical component that fostered rapid technological advancement. In fact, the growth was so dramatic that by 1965 Gordon Moore developed Moore's law as a way to predict the increase in computer power [18]. The law states that computer power will double at a constant cost roughly every two years.

Moore's predictions have held true for decades, but conventional fabrication techniques are running up against a wall. Transistor sizes are becoming much smaller (the current state of the art is approximately 4–5 nanometers) as devices are made up of more and more of them with every technological advancement with the goal of making devices thinner and more efficient. However, with silicon technology, there is a fundamental limit on how much size of transistors could be scaled down and that is the size of a silicon atom which is approximately 0.2 nm. Further, quantum effects, which do not generally affect our macroscopic world, become non-negligible at the nanoscale level. This has been a motivation to research a new paradigm: quantum computing and quantum information. Quantum computing tries to exploit the strangeness of quantum physics to develop devices and computers that may provide speed and advantage for solving difficult and interesting problems, and transform cybersecurity. These problems include simulations of complicated molecules which can help in the synthesis of new drugs or fertilizers.

Development of Quantum Computing

The idea of building a quantum computer was first proposed in 1980 at the First Conference on the Physics of Computation at MIT.

Renowned physicists Richard Feynman and Paul A. Benioff gave talks about how a computer can operate under the laws of quantum mechanics and why a quantum computer would be the best way to simulate quantum systems [19]. Feynman proposed a high-level model for building a quantum computer. These new ideas inspired other scientists to build on these early proposals. David Deutsch described the first universal quantum computer and later proposed a computational problem that could be solved efficiently on a quantum computer but not with a classical computer [20].

Before the discovery of Shor's Algorithm in 1994 by computer scientist Peter Shor, research in this new field was still slow. After his discovery, the United States government held a workshop on quantum computing, and proposals for realizing quantum computers and experimental work began. What was so groundbreaking about Shor's Algorithm? The algorithm showed that prime factorization for large integers and the discrete logarithm problem could be solved efficiently on quantum computers [21]. Again, why is this relevant? Well, theoretically, this meant that many existing crypto-systems in the world could be broken if a quantum computer existed! Widely used crypto-systems like RSA rely on prime factorization of large numbers being very difficult, implying that you cannot break the code in a human lifetime. You can start to understand why even the theoretical idea that crypto-systems could be broken would spark such an immediate interest. It marked the start of an important and exciting race (more like a marathon) of who could build the first quantum computer and essentially save themselves from the potential security threat.

Of course, these immediate actions from different sides would make one think that there was an immediate threat, but the truth is we are still far away from realizing a quantum computer capable of implementing Shor's algorithm to such a large scale. Nonetheless, this has not receded the interests of researchers, industry leaders and world governments.

Goal and Structure of This Book

The goal of this book is to be a starter guide to beginners interested in quantum information theory. Oftentimes, the *introductory* content in this field requires too many prerequisites. This book is designed to be self-contained so that the reader does not need to reference other materials to understand the contents of the book. But, we hope that this book inspires you to seek out more content.

The book is organized to give the necessary background to understand some of the basic principles within the subject of quantum computing. Qubits, the fundamental units of quantum information, are described as complex vectors. To understand and operate on complex vectors, we provide introduction to complex numbers and linear algebra. Measurement and extraction of information from qubits require knowledge of probability theory since measurements are probabilistic. The theory that combines the mathematics and the physical interpretation to understand qubits lies in quantum mechanics.

In the first and second chapters of the book, we give the mathematical background or the language of quantum mechanics which includes discussion and examples of working with complex numbers, combinatorics and probability theory, and linear algebra. In the third chapter, we introduce quantum mechanics and discuss systems that can be used as qubits, such as photon polarization and spins in magnetic fields. In the fourth and fifth chapters, we discuss single qubits and how they can be controlled and measured, and examples showing their possible applications. In the sixth chapter, we discuss two qubits, the nature of entanglement, and what is possible with many qubits. The application discussed is quantum teleportation and why it is of interest. Finally, chapter seven discusses the experimental implementation of quantum computers, the current state of the art, and the relevant concerns researchers face in trying to develop this technology.

Chapter 1

Preliminary Math Tools

To begin our study of quantum information, let's first review some mathematical concepts and tools that are fundamental to the language that describes quantum information and will be seen throughout the book. The first of these is complex numbers. The language of quantum mechanics inherently works in a complex vector space, so **qubits**, the fundamental unit of quantum information, are represented by complex numbers. The second tool inherent to the language and interpretation of quantum mechanics, is probability theory. In this chapter, we cover the basics of random variables and probability distributions which will be necessary to understand later chapters of the book.

1.1 Complex Numbers

We know that the "normal" numbers we work with are numbers that live in a one-dimensional line, called the *real* number line, as shown in Fig. 1.1. These are numbers that we typically encounter, and they could be integers, rational numbers which are defined as a ratio of two integers (e.g. $\frac{2}{3}$), or irrational numbers that cannot be expressed as a ratio of two integers. Irrational numbers include mathematical constants such as $\pi \approx 3.14159\ldots$, which is defined as the ratio of a circle's circumference to its diameter. Another famous mathematical constant is $e \approx 2.71828\ldots$, known as Euler's number. This number comes up in the study of compounded interest, probability, and complex numbers.

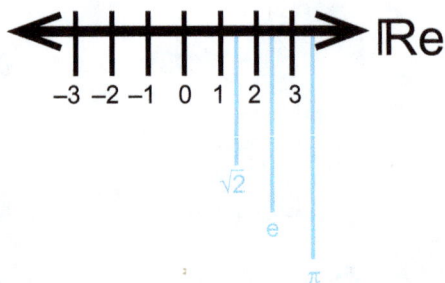

Figure 1.1. Real number line, including integers, rational and irrational numbers.

The real number line does not include the result you would get if you calculated the square root of a negative number, for example, $\sqrt{-1}$. It turns out that the result of calculating the square root of negative numbers plays an important role in physics. To do so, the real number line is not sufficient and a new dimension or axis must be added to include the results of taking the square root of negative numbers. This new axis includes what we call **imaginary numbers**, which are labeled by the real number times i, for example, $i, 2i, 3i, \ldots$. Specifically, we define one unit along the imaginary axis to be

$$\sqrt{-1} = i$$

which means that

$$i^2 = -1$$
$$i^3 = -i$$
$$i^4 = 1$$

and so on.

Adding this new axis generalizes what numbers are. It tells us that numbers can have real *and* imaginary components. This new combination is known as a **complex number**. So, complex numbers will now live on a *complex plane* composed of a real number axis and an imaginary number axis, as shown in Fig. 1.2. Any point on the plane is a complex number.

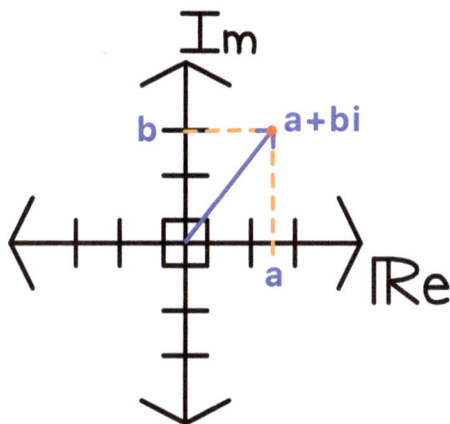

Figure 1.2. Definition of complex plane consisting of the imaginary and real axes.

1.1.1 Standard Representation

Here are some properties of complex numbers, $z_1 = a + bi$, $z_2 = c + di$, where a and c are the real parts and b and d are the imaginary parts, respectively:

1. **Addition:** $z_1 + z_2 = (a + bi) + (c + di) = (a + c) + (b + d)i$.
2. **Multiplication:** $z_1 z_2 = (a + bi)(c + di) = a(c + di) + bi(c + di) = ac - bd + i(ad + bc)$.
3. **Conjugate:** For $z_1 = a + bi$, the conjugate is $z_1^* = a - bi$.
   The product of a complex number with its conjugate yields $z_1 z_1^* = (a + bi)(a - bi) = a^2 + b^2$.

We can also define the angle, θ, with respect to the real number axis, for a line connecting the origin and the point where the complex number is located. Then, we may use trigonometric functions to describe complex numbers as you would on a regular Cartesian plane.[1] To do so, we also need to know the distance, D, between the origin and the point where the complex number is located. This quantity is also known as the **magnitude** of a complex number, z, and is denoted as $|z|$. We can calculate this quantity by using the distance formula:

$$|z| = D = \sqrt{a^2 + b^2} = \sqrt{zz^*}$$

---

[1]Refer to Appendix chapter C for a review of the trigonometric functions.

So, we can now see the relationship between the complex number and its conjugate. Then, θ can be written as

$$\cos\theta = \frac{a}{D}$$

$$\sin\theta = \frac{b}{D}$$

$$\tan\theta = \frac{b}{a}$$

Examples

1. Let's say we have two complex numbers, $z_1 = 4 + 3i$ and $z_2 = 5 - 2i$. We may plot the numbers on the complex plane to visualize them in Fig. 1.3.

   As we can see, if we add these two complex numbers, we are adding the real components and the complex components to each other, respectively, so that $z_3 = z_1 + z_2 = (4 + 5) + (3 - 2)i = 9 + i$. Let's also use trigonometry to calculate the relative angle

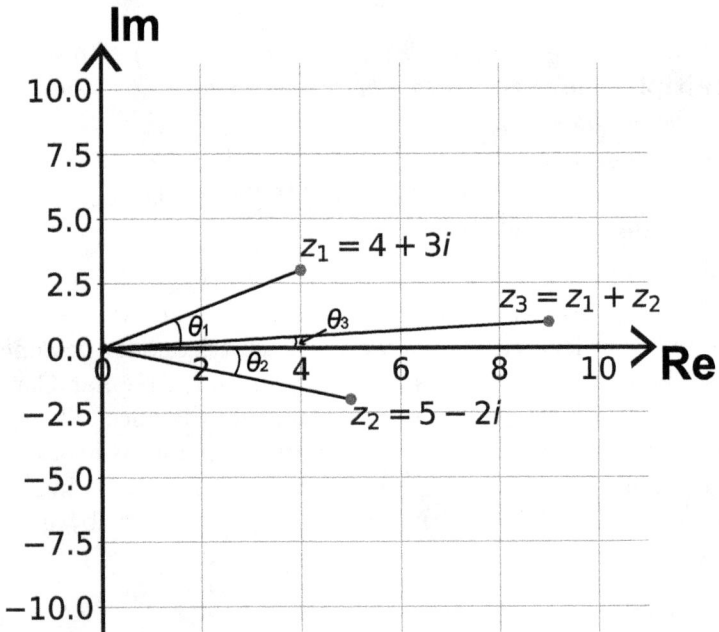

Figure 1.3.   Complex plane plot of example 1.

between the real axis and line that connects the complex numbers and origin. For z_1, we label angle θ_1 from the real axis in the counterclockwise direction:

$$\tan \theta_1 = \frac{3}{4} \longrightarrow \theta_1 = \arctan\left(\frac{3}{4}\right) \approx 37°$$

For z_2, we label angle θ_2 from the real axis in the clockwise direction:

$$\tan \theta_2 = \frac{-2}{5} \longrightarrow \theta_2 = \arctan\left(\frac{-2}{5}\right) \approx -22°$$

Lastly, for z_3, we label angle θ_3 from the real axis in the counterclockwise direction:

$$\tan \theta_3 = \frac{1}{9} \longrightarrow \theta_3 = \arctan\left(\frac{1}{9}\right) \approx 6°$$

2. Let's try to multiply z_1 by the imaginary number $2i$. Algebraically, we may compute this easily now that we know the properties of complex numbers:

$$2i \cdot (4 + 3i) = 8i + 2 \cdot 3i^2 = 8i - 6$$

What does this mean geometrically? The plot is shown in Fig. 1.4.

While multiplying the complex number by the imaginary number $2i$, we observe that the magnitude of z_1 is scaled by a factor of 2. More interestingly, it rotated z_1 by 90° counterclockwise. If we had multiplied by $-2i$, there still would have been a scaling by a factor of 2, but the rotation would have been 90° clockwise instead.

3. Now, let's multiply two complex numbers, $p_1 = 3 + i$ and $p_2 = 4 + 2i$, we would obtain the following result:

$$p_3 = p_1 \cdot p_2 = 3(4 + 2i) + i(4 + 2i) = 12 + 6i + 4i + 2i^2$$
$$= (12 - 2) + i(6 + 4) = 10 + 10i$$

Let's visualize this by plotting the complex numbers on the complex plane in Fig. 1.5.

We can tell that the resulting complex number, p_3, is stretched and rotated by some amount. In fact, the amount that p_1 is rotated

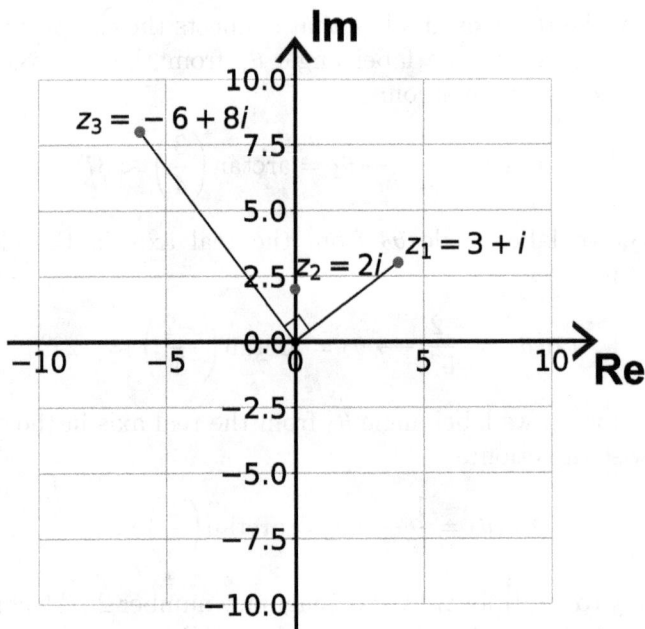

Figure 1.4.   Complex plane plot for example 2.

by to get to p_3 is exactly α_2, the relative angle that defines the line connecting point p_2 with the origin. Let's compute α_1 and α_2 to see if they add up to α_3:

$$\tan \alpha_1 = \frac{1}{3} \longrightarrow \alpha_1 = \arctan\left(\frac{1}{3}\right) \approx 18.4°$$

$$\tan \alpha_2 = \frac{2}{4} \longrightarrow \alpha_1 = \arctan\left(\frac{2}{4}\right) \approx 26.6°$$

$$\tan \alpha_3 = \frac{10}{10} = 1 \longrightarrow \alpha_1 = \arctan(1) = 45°$$

Indeed, $\alpha_1 + \alpha_2 = \alpha_3$.

Now, let's see how the magnitude of p_1 changed when p_1 and p_2 were multiplied. We compute the distances from the origin or the magnitudes of p_1, p_2, and p_3 and denote them as $|p_1|$ $|p_2|$, and $|p_3|$,

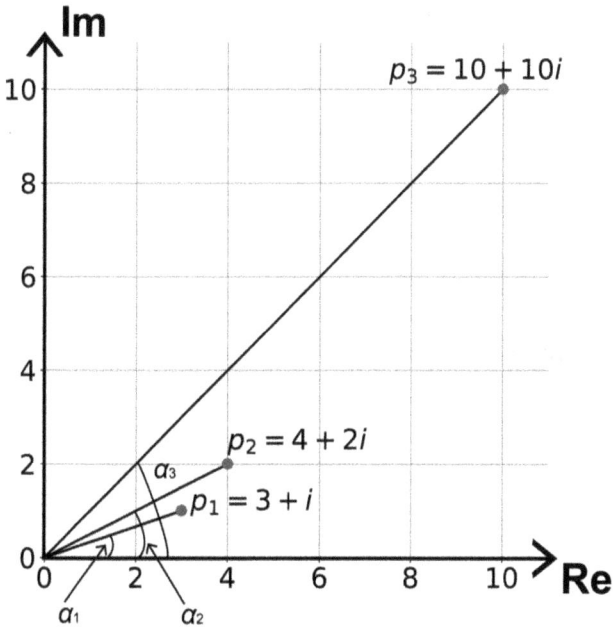

Figure 1.5. Complex plane plot for example 3.

respectively:

$$|p_1| = \sqrt{3^2 + 1^2} = \sqrt{10}$$
$$|p_2| = \sqrt{4^2 + 2^2} = \sqrt{20}$$
$$|p_3| = \sqrt{10^2 + 10^2} = \sqrt{200} = \sqrt{10}\sqrt{20}$$

As in the previous example, the magnitude of p_1 was scaled by the magnitude of p_2 to produce p_3.

4. Lastly, let's compute the result of dividing a complex number by an imaginary number:

$$\frac{3 + 12i}{2i}$$

As a convention, having complex numbers in the denominator is not preferred, so we need to multiply both top and bottom of the fraction by the conjugate of the term in the denominator so that we can have a real number in the denominator:

$$\frac{3+12i}{2i} \cdot \frac{-2i}{-2i} = \frac{-(3\cdot2)i - (12\cdot2)i^2}{-(2\cdot2)i^2} = \frac{-6i+24}{4} = 6 - \frac{3}{2}i$$

In this case, having divided by $2i$ ended up being as if we multiplied by $-2i$, or rotating clockwise by $90°$, and scaled by $\frac{1}{2}$. Indeed, the magnitude of $3+12i$ is ≈ 12.4 and magnitude for the final answer $6 - \frac{3}{2}i$ is ≈ 6.2.

1.1.2   Polar Representation

We saw in example 2 that multiplying a complex number by an imaginary number leads to a $90°$ rotation. So, if we were to take the number 1 and multiply it by i and continue multiplying the resulting values by i, we would have

$$1 \cdot i = i$$
$$i \cdot i = i^2 = -1$$
$$-1 \cdot i = -i$$
$$-i \cdot i = -i^2 = 1$$

As we can see, the effect of multiplying by an imaginary number is periodic because we returned back 1 after 4 multiplications by i. In fact, what we did was go around a circle. We started at $0°$, then rotated counterclockwise, $90°$, $180°$, $270°$ (or $90°$ rotation *clockwise*), and $360°$ (or back to $0°$). Thinking about complex numbers in this way allows us to represent them in *polar form*, where we are defining the magnitude of the complex number, or the distance from the origin, and the angle it is rotated by relative to the real axis.

A compact way to represent this rotation around a circle is using Euler's formula:

$$e^{i\theta} = \cos\theta + i\sin\theta$$

There are direct similarities with this representation to the unit circle from trigonometry which is defined on the xy-plane, as can be seen in Fig. 1.6. The x-axis now corresponds to the real component of the complex number and y-axis to the imaginary component. The angle θ can be computed using trigonometry, as we did in the previous section.

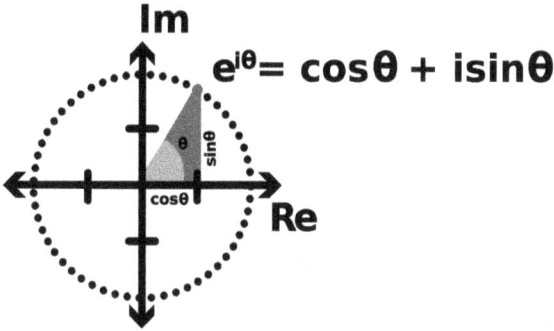

Figure 1.6. Unit circle rotations.

Any complex number that we represented using the standard form can be rewritten using Euler's formula to polar form. While Euler's formula represents the unit circle, if we scale a unit circle by any other number, then we can obtain circles with different radii. The thought process behind this is to multiply the number $e^{i\theta}$ by the magnitude $(|z| = \sqrt{z^*z})$ of that complex number, so we have a complex number that falls on the circle of radius $|z|^2$:

$$z = a + bi = \sqrt{z^*z}(\cos\theta + i\sin\theta) = |z|\, e^{i\theta}$$

Examples

1. Let's say we have a complex number $z = 4+3i$ and we want to represent it in polar form. We need to compute the magnitude of z,

$$|z| = \sqrt{4^2 + 3^2} = \sqrt{25} = 5$$

and the relative angle from the real axis:

$$\varphi = \arctan\left(\frac{3}{4}\right) \approx 37° \text{ or } 0.64 \text{ rad}$$

So, now, we have found that our complex number falls on a circle whose radius is 5 and is located at an angle $\varphi \approx 0.64$ rad from the real axis. Let's visualize this result in Fig. 1.7.

---

[2] As per convention, the angle, φ, that we put into Euler's formula is generally presented in radians instead of degrees.

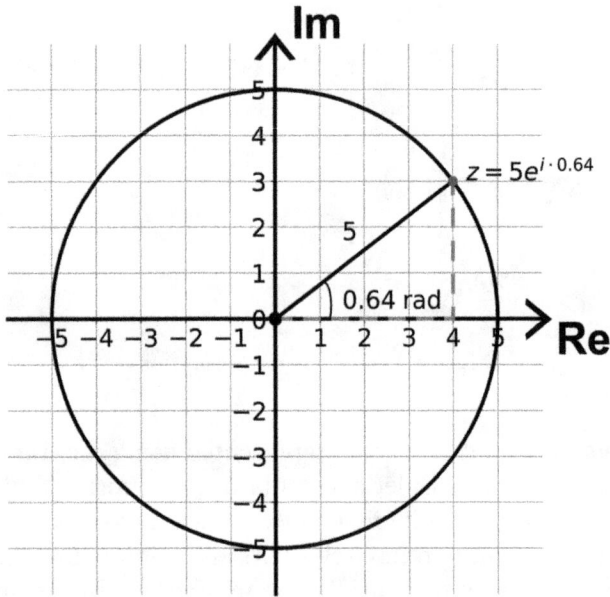

Figure 1.7.   Complex number represented in polar form for Example 1.

2. Let's redo the examples of multiplying two complex numbers $p_1 = 3 + i$ and $p_2 = 4 + 2i$ by first converting them to the polar representation:

$$|p_1| = \sqrt{3^2 + 1^2} = \sqrt{10} \quad \varphi_1 = \arctan\left(\frac{1}{3}\right) \approx 0.32 \text{ rad}$$

$$\longrightarrow p_1 = \sqrt{10}\, e^{i0.32}$$

$$|p_2| = \sqrt{2^2 + 4^2} = \sqrt{20} \quad \varphi_2 = \arctan\left(\frac{2}{4}\right) \approx 0.46 \text{ rad}$$

$$\longrightarrow p_2 = \sqrt{20}\, e^{i0.46}$$

Okay, now we are ready to multiply[3]:

$$p_3 = p_1 \cdot p_2 = \left(\sqrt{10}e^{i0.32}\right) \cdot \left(\sqrt{20}e^{i0.46}\right) = \sqrt{10} \cdot \sqrt{20}\, e^{i0.32} \cdot e^{i0.46}$$

$$= \sqrt{200}\, e^{i(0.32+0.46)} = \sqrt{200}\, e^{i0.78}$$

---

[3]Refer to Appendix chapter C for a review of multiplying exponentials.

Let's convert back to the standard notation to ensure we got the same answer as in the previous section[4]:

$$p_3 = \sqrt{200}\, e^{i0.78} = \sqrt{200}(\cos 0.78 + i\sin 0.78)$$

$$\approx 14.14(0.71 + i0.70) \approx 10 + 10i$$

Also, if we convert 0.78 rad to degrees, we get

$$0.78 \cdot \frac{180°}{\pi} \approx 45°$$

So, with the polar representation, it is much clearer that when we multiply two complex numbers, their angles actually add. This is the benefit of using the polar representation.

3. Lastly, let us do the last example from the previous section

$$\frac{3 + 12i}{2i}$$

using the polar representation. First, let's write the complex number in the numerator (let's label it n) and the imaginary number in the denominator (let's label it d) in polar form:

$$|n| = \sqrt{3^2 + 12^2} = \sqrt{153} \quad \varphi = \arctan\left(\frac{12}{3}\right) \approx 1.33 \text{ rad}$$

$$\longrightarrow n = \sqrt{153}\, e^{i1.33}$$

$$|d| = \sqrt{2^2} = 2 \quad \varphi = 90° = \frac{\pi}{2} \approx 1.57$$

$$\longrightarrow d = 2\, e^{i1.57}$$

So, using the rules of dividing exponentials,

$$\frac{n}{d} = \frac{\sqrt{153}\, e^{i1.33}}{2\, e^{i1.57}} = \frac{\sqrt{153}}{2} \cdot \left(e^{i(1.33-1.57)}\right) \approx 6.18\, e^{-i0.24}$$

We can check that the answers are the same:

$$6.18\, e^{-i0.24} = 6.18(\cos(-0.24) + i\sin(-0.24))$$

$$\approx 6.18(0.97 - i0.24) \approx 6 - i1.5$$

In the following chapters, we will see both representations of complex numbers used to write down qubit states.

---

[4]Since we rounded the values for some of the angles, the answer is not exact but very close.

1.2   Probability

1.2.1   Introduction

In a few words, probability is a branch of mathematics that provides numerical descriptions on the likelihood of certain events. The simplest probability experiment is the coin toss. We know that for a fair coin, the probability of getting heads or tails is $\frac{1}{2}$. This means that each event is equally likely because the *sample space*, consisting of all the possible outcomes of the experiment, is {Head, Tail}. If we two coins at the same time, then the sample space would be {HH, HT, TH, TT}. We can visually understand this by making a table, where the first row represents the outcomes for one of the coins and the first column represents the outcomes for the other coin:

	H	T
H	HH	HT
T	TH	TT

A key point to make about this kind of experiment is that we are assuming that events are independent of each other, implying that if you get heads in one trial, it does not affect the result in the next trial.

So, if we define event A to be getting heads and event B to be tails, then we can mathematically express this as

$$P(A) = \frac{1}{2}$$

$$P(B) = \frac{1}{2}$$

An important point is that events A and B cannot occur at the same time, implying that you cannot get both heads and tails in any one trial. Events that cannot happen together in any given trial are *mutually exclusive* or *disjoint* thus:

$$P(A \text{ and } B) = 0$$

But, the probability of getting heads or tails is

$$P(A \text{ or } B) = \frac{1}{2} + \frac{1}{2} = 1$$

This means, in any trial, you will get heads or tails with 100% chance.

This coin toss experiment is an example of a **Bernoulli trial**. A Bernoulli trial is a randomized experiment with exactly two possible outcomes. In such a trial, you define a probability of "success" and a probability of "failure." This idea appears in many real-world applications. For example, in epidemiology, one is concerned about infection rates, where for any given person in the population, there is some probability, p, that they are infected and probability, $q = 1 - p$, that they are not infected.

Rules of Probability

1. For all events, E, in the sample space, S, the probability of that event of occurring must lie between 0 and 1.
2. The probability of the sample space, S, must equal 1. This is because the sample space contains all the possible outcomes of a random experiment and whatever event occurs belongs to S, so the probability of the sample space "occurring" is always 1.
3. For several disjoint events, $E_1, E_2, E_3, ...$, the probabilities of getting any of those events add[5]:

$$\mathbb{P}(E_1 \text{ or } E_2 \text{ or } E_3 \text{ or } ...) = \mathbb{P}(E_1) + \mathbb{P}(E_2) + \mathbb{P}(E_3) + \cdots$$

Checkpoint Exercise

If we roll 2 tetrahedron dice (each die has only 4 sides labeled 1,2,3,4), what is the probability that the sum of the the 2 outcomes is 5?

1.2.2 Combinatorics

Combinatorics is concerned with counting. We will cover what is necessary to understand the common probability distributions we will encounter with qubits. First, let us introduce the factorial operation:

$$n! = n \cdot (n-1) \cdot (n-2) \cdot (n-3) \cdot \cdots \cdot 1$$

where n is a non-negative integer.[6]

---

[5]The "or" statement is to denote that in the case of disjoint events you can have one event *or* the other, etc. and the probabilities will add because the events cannot occur at the same time.
[6]$0! = 1$.

You may already be familiar with the **basic counting principle.** For example, if we have 4 different colored shirts and 5 different colored pants then for one shirt there are 5 different possible colors for pants, so the number of possible outfits is $4 \times 5 = 20$.

Now, what if we wanted to consider in how many ways we could arrange the shirts and pants in our closet if we want 3 unique outfits? Well, we can think about the shirts and pants separately. For the first day, we have 4 options for shirts, the second day, we have 3 options (because we want unique outfits), and the third day, we have 2 options. So, the number of ways to wear the shirts is $4 \times 3 \times 2 = 24$. Similarly, for pants, 5 options for first day, 4 options for second day, and 3 options for the third day. So, the number of ways to wear the pants is $5 \times 4 \times 3 = 60$. The total number of ways to arrange the outfits is $24 \times 60 = 1440$.

Now, in how many ways could we choose 2 shirts and 2 pairs of pants to pack for a two-day trip? In this case, the order in which we count the shirts and pants does not matter. We are choosing 2 shirts from our pile of 4 shirts, so there are 4 options for first choice and 3 for the second choice. We know there are only 2 ways to arrange any 2 shirts since we don't care about the order, the number of combinations is $\frac{4 \times 3}{2} = 6$. Similarly, for pants, we choose 2 from a pile of 5 pairs of pants, where there are 2 ways to arrange 2 pants, so the number of combinations is $\frac{5 \times 4}{2} = 10$. The total number of ways to pack 2 shirts and 2 pairs of pants in this case would be $6 \times 10 = 60$ ways.

While these problems are simple, if the number of objects we are dealing with is large, it's best to use a formula. For the problem where we are concerned with how many ways we could arrange 3 unique outfits, we can simply use the **permutations** formula:

$$ _nP_k = \frac{n!}{(n-k)!} $$

This formula tells us the number of arrangements or permutations of k objects from a set of n. So, for our problem, we wanted to choose 3 unique shirts from a set of 4: $_4P_3 = \frac{4!}{(4-3)!} = 24$. Similarly, we wanted 3 unique pairs of pants from a set of 5: $_5P_3 = \frac{5!}{(5-3)!} = 60$.

For the second case, where we wanted to know how many ways we could choose our shirts and pants for the day trip, we were not

concerned about the order that we were choosing. We could use the
combinations formula in this case:

$$_nC_k = \binom{n}{k} = \frac{n!}{k!(n-k)!}$$

The modification here from the permutations formula is that we are
also dividing by the number of ways we can arrange k objects because
we don't care about the order they are in unlike permutations. So, if
we want the number of ways to choose 2 shirts from a set 4, $_4C_2 = \frac{4!}{2!(4-2)!} = 6$. Similarly, if we want 2 pairs of pants from a set of 5,
$_5C_2 = \frac{5!}{2!(5-2)!} = 10$.

Checkpoint Exercises

1. How many unique license plate numbers with 3 letters followed by
   2 numbers are possible? (*Hint*: There are 26 letters in the alphabet
   and numbers may be from 0 to 9.)
2. In how many ways can you arrange 3 math textbooks, 1 chemistry textbook, and 5 physics textbooks (because you *love* physics
   obviously) so that the all the math books are stacked up next each
   other, all the chemistry books are stacked up next to each other,
   and all the physics books are stacked up next to each other?
3. Suppose we have two basketballs teams, *Team 1* and *Team 2*,
   where *Team 1* has 5 players and *Team 2* has 6 players.
   a. How many ways can we arrange the players if all players in
      Team 1 stand together?
   b. Suppose you want to choose 3 players, which either have to be
      all from *Team 1* or all from *Team 2*?

1.2.3 Discrete Random Variables

As we discussed previously, when dealing with random events, it is
useful to define a sample space, S, that consists of all the possible
outcomes of an experiment. Further, we can generalize the way we
consider these outcomes by defining a useful quantity known as a
random variable. This variable can be *discrete*, meaning it has specific
values over some range, or *continuous*, where it can take any value
over a specified range. For the purposes of this book, we will focus
on discrete random variables.

Table 1.1.   Probabilities
for random variable Z.

Z	$\mathbb{P}(Z = z)$
0	$\left(\dfrac{5}{6}\right)^3$
1	$3\left(\dfrac{1}{6}\right)\left(\dfrac{5}{6}\right)^2$
2	$3\left(\dfrac{1}{6}\right)^2\left(\dfrac{5}{6}\right)$
3	$\left(\dfrac{1}{6}\right)^3$

In fact, a random variable is a function of the random outcome. For example, let's consider rolling a die one time, and define X to be a random variable for rolling the dice once. We know that X could take on the values 1, 2, 3, 4, 5, or 6 (the outcomes of rolling the die) which are all equally likely to be $\frac{1}{6}$. Now, let's suppose we roll a die once and define a random variable Y to be 1 for obtaining an odd outcome (1,3,5) and 0 for even outcomes. Both even and odd outcomes are equally likely so that we obtain $\frac{1}{2}$ of the time odd and $\frac{1}{2}$ even.

Let's do a more complicated example: let random variable Z denote the number of 4s we get when we roll a die three times. Well, in this case, Z can take on the values 0, 1, 2, 3 because there is a chance we will not roll a four or we roll a four just once, or twice, or all three times. It is best to make a table and assign a probability to each possible outcome as done in Table 1.1.

Let's understand why those are the correct probabilities. For $Z = 0$, we would get $\left(\frac{5}{6}\right)^3$ because $\frac{5}{6}$ is the probability of NOT getting a 4, but because each dice roll is independent, the probabilities multiply. For $Z = 1$, we would have one occurrence of 4 (with probability $\frac{1}{6}$), but this can happen in $\binom{3}{1} = 3$ ways (on the first roll, second roll, or third roll) which is why we have "3" in front. The same idea follows for $Z = 2$ where $\binom{3}{2} = 3$. For $Z = 3$, we would get 4s for all three rolls where each roll is independent. So, the distribution we have in this case is

$$\mathbb{P}(Z = k) = \binom{3}{k} p^k (1 - p)^{n-k}$$

This is known as the binomial distribution, where probability p represents the probability of "success," in this case getting a 4, and we are repeating the experiment a total of 3 times. So, the binomial distribution may be considered as repeating n Bernoulli trials.

As you can see, defining random variables is a very versatile way of working with probabilities and allows for generalizations to *probability distributions*. These are mathematical functions that give the probabilities of different outcomes for a certain experiment and the input to these functions are the set of values which a specified random variable can take.

Checkpoint Exercises

1. Suppose we toss a fair coin 2 times. Let random variable S denote the number of heads that appear.

   a. What are the values that S can take?
   b. What are the probabilities associated with each of those values?

2. Eddie has been practicing his free throws and has a probability of 0.8 of making it successfully. Suppose that each free throw is independent of the other. If Eddie attempts 8 free throws, what is the probability that he makes at least 3 of them successfully?
3. Pfizer conducted a study on a new version of their COVID-19 vaccine. Their study showed that the vaccine was 80% effective in preventing COVID-19. What is the probability that of 6 randomly selected patients, only 2 of them contract COVID-19 after having received the vaccine?

1.2.4   Calculating Expected Values

You may wonder where we are heading with probabilities. Well, we can use these ideas to calculate some useful quantities like the *expected value* of an experiment. The expected value of a random

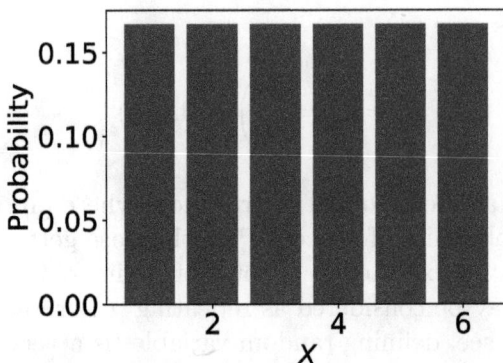

Figure 1.8.　Probability mass function for random variable X.

variable, X,[7] is defined as

$$\mathbb{E}X = \sum_x x p_X(x)$$

where $p_X(x)$ is the called the probability mass function (PMF) and assigns a probability to each possible outcome of some experiment. Let's go back to the example of die with the defined random variable, X. The probability mass function can be defined as follows:

$$p_X(x) = \begin{cases} \frac{1}{6}, & 1 \le x \le 6 \\ 0, & x > 6 \end{cases}$$

Then we can calculate the expected value to be

$$\mathbb{E}X = 1 \cdot \frac{1}{6} + 2 \cdot \frac{1}{6} + 3 \cdot \frac{1}{6} + 4 \cdot \frac{1}{6} + 5 \cdot \frac{1}{6} + 6 \cdot \frac{1}{6} = \frac{7}{2}$$

The expected value or the *weighted* average of the random variable X is 3.5. We can visually identify the expected value as the "center of mass" of the distribution. The distribution, p_X, is shown in Fig. 1.8.

If we were considering the random variable Z we worked with in the previous section, the corresponding PMF may be graphed as shown in Fig. 1.9.

Then, the expected value is

$$\mathbb{E}Z = 0 \cdot \left(\frac{5}{6}\right)^3 + 1 \cdot 3 \left(\frac{1}{6}\right)\left(\frac{5}{6}\right)^2 + 2 \cdot 3 \left(\frac{1}{6}\right)^2 \left(\frac{5}{6}\right) + 3 \cdot \left(\frac{1}{6}\right)^3 = \frac{1}{2}$$

---

[7]Random variable, X, may take values $X = 0, 1, 2, 3, ..., x$, as needed for the experiment we are describing.

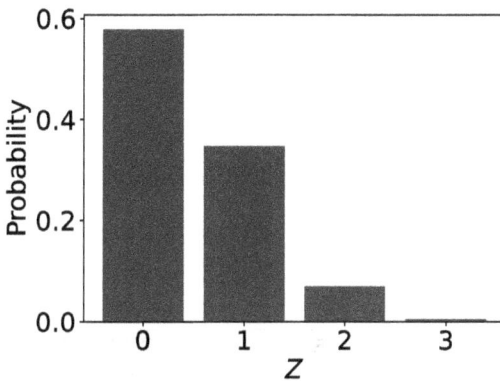

Figure 1.9. Probability mass function for random variable Z.

Checkpoint Exercises

1. Refer back to your solution of checkpoint exercise #1 in the previous section. What is the expected value of the experiment?
2. Suppose that we have a random variable Y that has the following PMF:

$$p_Y(y) = \begin{cases} 1/8, & y = 0 \\ 3/8, & 1 \leq y \leq 2 \\ 1/8, & y = 3 \end{cases}$$

    a. Plot the distribution by hand or using a graphing tool.
    b. Compute the expected value of the distribution.

1.2.5 Law of Large Numbers

So far, we have discussed some simple probability experiments like the coin toss and die roll. We understand that if we toss a coin, we have a probability of $\frac{1}{2}$ to get heads or tails, and that if we roll a die, we have a probability of $\frac{1}{6}$ to get 1, 2, 3, 4, 5, or 6. These should both be uniform PMFs.

Let's suppose that we toss that coin 25 times. We will not get an equal number of heads or tails. We will get either more heads or more tails. In fact, this variation will always exist, but as we toss the coin N times, where N is a very large number that conceptually approaches infinity, we see the following behavior plotted in Fig. 1.10.

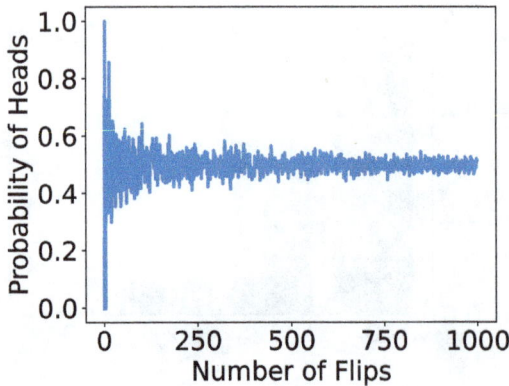

Figure 1.10.   Number of coin tosses vs. probability of getting heads (or tails).

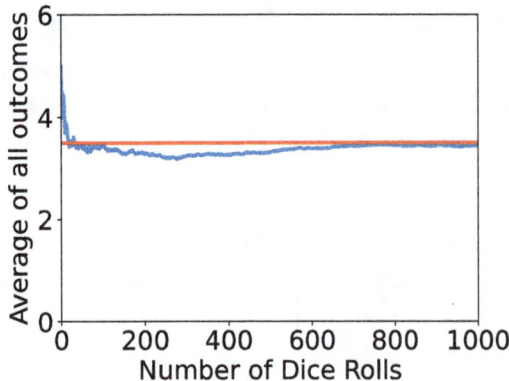

Figure 1.11.   Number of dice rolls vs. average value.

As the number of flips becomes very large, we approach the expected value for the coin toss experiment 0.5. Likewise, in the case of rolling a fair die, Fig. 1.11 shows the behavior. Once again, we can see that the average value over the many trials approaches the expected value of 3.5 that we calculated previously.

This is known as the *law of large numbers* where the averaged results from many trials approaches the expected value. It is significant for guaranteeing long-term stability for averages of random events. This idea has many applications in probability, statistics, finance, and physics, so we will see it in action in the following section.

1.2.6 Information Entropy

In physics, the concept of entropy arose in thermodynamics as a measure of the randomness or uncertainty in a physical system. For example, how gas particles disperse to fill the volume of a room. In 1870, Ludwig Boltzmann developed the statistical definition of entropy by analyzing the behavior of the microscopic constituents of the system. He defines entropy to be the logarithmic measure of the number of microstates in a system that has a significant probability of being occupied, defined as follows[8] [22]:

$$S = -k_B \sum_i p_i \ln p_i$$

where p_i is the probability that the system is in the ith state and this probability depends on the distribution of the states and $k_B \approx 1.38 \cdot 10^{-23}$ J/K is the Boltzmann constant.

Claude Shannon took the concept of entropy from physics and applied it to information theory in 1948 as a way to represent the "degree of surprises" that the transfer of some information can bring [23]. Shannon entropy, H, is defined in a way similar to the previous definition:

$$H(X) = -\sum_x p(x) \log_b p(x)$$

where x represents the values that a random variable X can take and b is the base of the logarithm which is 2. If you knew exactly what the outcome of the message would be, then the Shannon entropy is 0. As an example, consider that we are tossing a coin which will give an equal outcome of head or tails depending on whether it is a fair coin or not. The more the coin is biased toward one particular outcome, the lower the entropy since the outcome will not surprise you as much as the outcome of a fair coin.

This entropy is especially important in information theory because it tells you the minimum number of bits that are needed to

---

[8]Here, the ln function is called the natural logarithm, which is a logarithm with base e, the base of the exponential form we saw previously to represent complex numbers in polar notation. Well, $e \approx 2.71828$ and is also called Euler's number. Refer to C for a review of logarithms.

convey a message, also known as data compression. If a sender uses fewer bits than this minimum, the message they are trying to send will be distorted. The concept of entropy comes up again in quantum information theory in the context of entanglement. Entropy can be used as a way to measure how much entanglement exists between two qubits.

Homework 1

Complex Numbers

Perform the indicated operation between the specified complex numbers in two ways:

1. in standard form and convert the final answer to polar,
2. in polar form and convert the final answer to standard form.

Plot the complex numbers and solution of the problems on a complex plane (by hand or using Python). Briefly describe how the complex numbers are rotated and scaled.

1. $z_1 = (1 - i)$, $z_2 = (6 - 5i)$, solve $z_1 \cdot z_2$.
2. $p_1 = 7 + i$, $p_2 = 2 + i$, solve $\frac{p_1}{p_2}$.
3. $d_1 = i$, $d_2 = 1 + i$, $d_3 = 1 + 3i$, solve $\frac{d_1 \cdot d_2}{d_3}$.

Probability Sample Spaces

Consider an urn containing 3 balls: 1 blue, 1 yellow, and 1 orange.

1. You take 1 ball from the urn for each turn and place the ball back in the urn each time before choosing the next ball. Write the sample space of the experiment, noting the blue ball as B, the yellow ball as Y, and the orange ball as O.
2. You take 1 ball from the urn and do not replace before drawing the second ball. Write the sample space of the experiment, noting the blue ball as B, the yellow ball as Y, and the orange ball as O.

Discrete Distributions

Suppose that an urn contains 5 blue, 4 yellow, and 3 orange balls. A ball is randomly drawn and replaced 5 times. Suppose the random variable X denotes the number of blue balls that are drawn from the urn.

1. What are the possible values that random variable X can take?
2. What are the probabilities associated with each value of X?
3. Plot the distribution.
4. What is the probability that at least 3 of the balls drawn are blue?

Calculating Expected Values

Continuing with the previous problem statement, where a ball is drawn and replaced 2 times. Suppose that \$2 can be earned for selecting a yellow ball and \$1 is lost for selecting a blue ball. Let the random variable E denote earnings.

1. What are the possible values of E?
2. What are the probabilities associated with each value of E?
3. What do you expect to earn?

Chapter 2

Basics of Linear Algebra

In order to start properly familiarizing ourselves with quantum computing, the first stepping stone is linear algebra which provides the mathematical tools to work with quantum information. The second stepping stone is quantum mechanics, the tool to understand the physical nature that governs quantum information. The following table shows some words that we encounter throughout this book. It highlights the connection between quantum mechanics and linear algebra.

Quantum Mechanics	Linear Algebra
Quantum state	Vector
Physical system	Hilbert space
Linear operators	Matrices
Measurable quantities	Hermitian matrices
Eigenstate	Eigenvector

Scalars are quantities that have no direction associated with them, for example, distance traveled, temperature, or mass. Vectors are objects that have both a magnitude and a direction. Vectors often show up in physics because they represent useful quantities, such as the position of an object, its velocity and acceleration, and the forces that are acting on it. For example, if a ball was rolling down an inclined plane, we can learn about its motion by measuring its speed (a scalar quantity) and observing that it moves *down* at the same angle as the incline which is the direction of the ball. Knowing the speed and direction of motion, we can define the velocity of the ball. The speed is the magnitude of the velocity and the direction is given by a unit vector, or a vector whose magnitude is one. In the off-chance that someone tries to block the ball from rolling down completely, then we say that there is a force acting on the ball in the direction opposing its motion, so the force is a vector quantity. One such opposing force would be the friction force the ball feels due to being in contact with the surface of the incline which slows the ball down. The point here is that the direction and magnitude of these physical quantities, such as velocity, acceleration, and force, are measurable.

A vector can be written in the row form, where its components are along one row as follows:

$$(x_1 \ldots x_n)$$

or in a column form:

$$\begin{pmatrix} x_1 \\ \vdots \\ x_n \end{pmatrix}$$

Vectors are used to represent a particular state of the physical system. For example, if a person moves from one place to another, we can give a number that tells us how much they moved and the direction they were moving along. This is otherwise known as *displacement*. In order for the numbers to mean something, we have to define the directions of movement that make sense to measure. For example, we may quantify how much someone moved in north and east. If the numbers are negative, then we may say they moved south or west instead. So, we think of north and east as being perpendicular directions that describe any kind of movement that someone could make on a flat plane. Alternatively, we can also measure along different directions like northeast and northwest. Basically, we just need to agree on two reference directions that are different from each other in order to describe the motion of someone on this flat plane.

Vectors live in **vector spaces** which are defined by the dimensions of the vectors in the space (e.g. 3D, 4D, or N-dimensional). The elements of the vector can be real (denoted $\mathbb{R}$) or complex (denoted $\mathbb{C}$) numbers. As such, when we talk about vector spaces, we provide the dimensionality and whether its vectors are real or complex. To say that you have a vector space is to say that there are certain operations that must be defined, such as addition and scalar multiplication. These operations must happen in such a way that they are "closed", meaning that the vector that results after performing the operation is in the same vector space (e.g. has the same dimension).

A matrix is a rectangular array of numbers arranged in rows and columns. A matrix "operates on" or "transforms" vectors to produce new vectors. These new vectors may have different dimensions or may keep the same dimension. An eigenvector is a special vector that when operated on by some particular matrix is only multiplied by a scalar called the eigenvalue. In other words, the vector would still point in the same direction, but it may be stretched or shrunk, depending on the eigenvalue.

We will see in the following chapters that in quantum mechanics, the quantum state (which represents the physical system) is expressed as a **complex vector** that lives in a complex vector space known as the Hilbert space. In this vector space, the length or *norm* of the complex vector can be defined, which is critical for physical states! Basically, finding the length of the vector removes its dimensionality and reduces the vector to a real-valued, positive scalar number. Matrices, or gates, which are linear transformations or linear operators, act on the quantum states to change their state. In particular, Hermitian matrices are special matrices that represent measurable quantities, such as position, momentum, and energy, etc. Lastly, the term *eigenstate* is the same as the eigenvector of these linear operators. The eigenvectors of the Hermitian matrices, in particular, are special because they represent the measurable outcomes.

2.1 Introduction to Linear Algebra

2.1.1 Euclidean Vectors

Vectors are mathematical objects that have both a magnitude[1] and a direction. Vectors are typically labeled with letters and can either

---

[1]Magnitude is the same thing as length, or distance from the origin, and it can be computed using the distance formula.

have an arrow on top or be bolded. For example, letter P, we would denote that P is a vector as

$$\vec{P} \text{ or } \mathbf{P}$$

One geometric representation of a vector is that it is a line with a given length which points in some direction denoted by an arrow. A common and intuitive way to describe vectors is through a Cartesian coordinate system with x, y, and z components. Figure 2.1 shows a 3D coordinate system which is defined by three mutually perpendicular or **mutually orthogonal** unit vectors, $\hat{i}$, $\hat{j}$, and $\hat{k}$. These unit vectors define the directions of the x, y, and z axes, respectively:

$$\hat{i} = \begin{pmatrix} 1 & 0 & 0 \end{pmatrix} \qquad \hat{j} = \begin{pmatrix} 0 & 1 & 0 \end{pmatrix} \qquad \hat{k} = \begin{pmatrix} 0 & 0 & 1 \end{pmatrix}$$

These vectors are called **unit vectors** because they have a length of one, so they will not affect the magnitude of the vector $\vec{P}$, but they define its direction. Using this definition, we write $\vec{P}$ as

$$\vec{P} = \begin{pmatrix} x & y & z \end{pmatrix} = x\,\hat{i} + y\,\hat{j} + z\,\hat{k}$$

This notation is telling us that we have moved x units along the x-axis or $\hat{i}$ direction (forward and backward), y units along the y-axis of $\hat{j}$

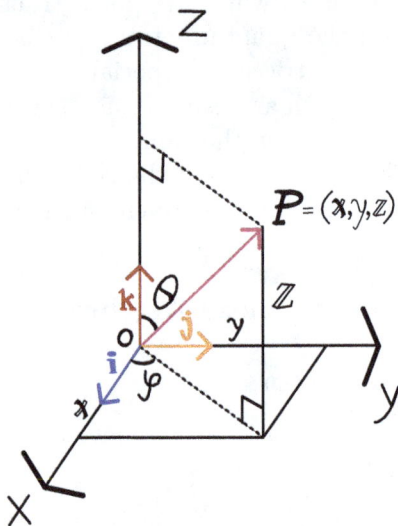

Figure 2.1.   Three-dimensional Cartesian coordinate system.

direction (left and right), and z units along the z axis, or $\hat{k}$ direction (up and down).

Vector Algebra

There are rules for doing algebra with vectors, including how to scale a vector by some amount and how to add and subtract two or more vectors together.

If we have a vector $\vec{P}$ and we scale it by a number a, then its length will be scaled by the number a, as shown in Fig. 2.2. For example, if $\vec{P} = 2\hat{i} + 4\hat{j} + 10\hat{k}$, and we multiplied by $a = \frac{1}{2}$, then we can distribute a to each of the vector components:

$$a \cdot \vec{P} = \frac{1}{2}(2\hat{i} + 4\hat{j} + 10\hat{k}) = 1\hat{i} + 2\hat{j} + 5\hat{k}$$

Now, $\vec{P}$ still points in the same direction, but it is $\frac{1}{2}$ its original length.

If we are adding two vectors $\vec{P}$ and $\vec{Q}$, we take the two vectors, put them tip to tail, and then draw a line connecting the end of the first vectors with the tip of the other. This is shown in Fig. 2.3(b). Alternatively, we can bring both vectors at the same initial point and complete the parallelogram. The diagonal of the parallelogram is the resultant vector $\vec{P} + \vec{Q}$.

As an example, suppose we have the vector $\vec{P} = 2\hat{i} + 4\hat{j} + 10\hat{k}$ and another vector $\vec{Q} = 3\hat{i} - 7\hat{j} - 1\hat{k}$, we wanted to compute the sum, $\vec{P} + \vec{Q}$. The sum of these two vectors will also be a vector because we can simply add the corresponding vector components to each other.

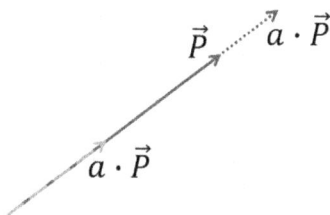

Figure 2.2. Scaling a vector $\vec{P}$ by a factor a. If $a > 1$, the magnitude of $\vec{P}$ will increase. If $0 < a < 1$, the magnitude of $\vec{P}$ will decrease. If $a < 0$, then the vector direction will flip by $180°$.

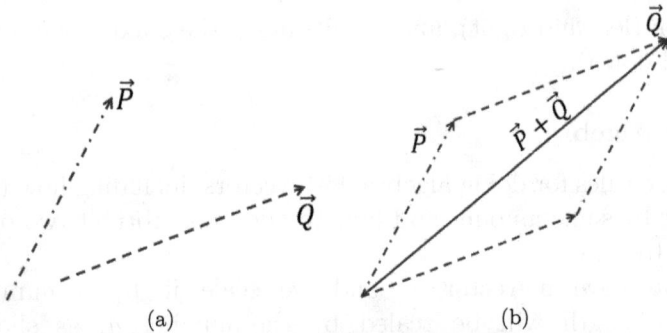

Figure 2.3. (a) Two vectors with different magnitudes pointing in different directions; (b) parallelogram method of adding vectors.

We will call the sum, $\vec{R}$, the resultant vector:

$$\vec{R} = \vec{P} + \vec{Q} = (2\hat{i} + 4\hat{j} + 10\hat{k}) + (3\hat{i} - 7\hat{j} - 1\hat{k})$$

$$= (2+3)\hat{i} + (4-7)\hat{j} + (10-1)\hat{k}$$

$$= 5\hat{i} - 3\hat{j} + 9\hat{k}$$

The only restriction to adding vectors is that the vectors must be of the same *dimension*. This is equivalent to saying that they have the same number of components or that they must be in the same vector space.

Dot Product

Now what if we also had another vector $\vec{Q} = x'\,\hat{i} + y'\,\hat{j} + z'\,\hat{k}$ and we wanted to project this vector onto $\vec{P}$ to identify how much or what percentage of $\vec{Q}$ is along $\vec{P}$? To do so, we consider that there is some angle, α, between the two vectors. As we can see in Fig. 2.4, the projection of $\vec{Q}$ on $\vec{P}$ is $|\vec{Q}| \cos \alpha$.

The dot product of $\vec{P}$ and $\vec{Q}$ is defined as

$$\vec{P} \cdot \vec{Q} = |\vec{P}||\vec{Q}| \cos \alpha$$

In general, to compute the dot product, we look at the actual vector components. Note that the result is a scalar and NOT a vector quantity. We can also think of the dot product as computing how much

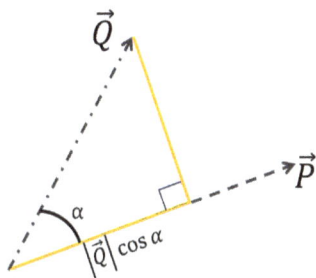

Figure 2.4. Representation of the dot product.

each component of the vectors overlap[2]:

$$\vec{P} \cdot \vec{Q} = (x\,\hat{i} + y\,\hat{j} + z\,\hat{k}) \cdot (x'\,\hat{i} + y'\,\hat{j} + z'\,\hat{k}) = x \cdot x' + y \cdot y' + z \cdot z'$$

The magnitude of the vector, $|\vec{P}|$, can be found using the distance formula between the point $(0,0,0)$ and (x,y,z):

$$|\vec{P}| = \sqrt{x^2 + y^2 + z^2}$$

Note here that the magnitude of a vector can also be defined in terms of the dot product:

$$|\vec{P}| = \sqrt{\vec{P} \cdot \vec{P}}$$

In other words, the length of the vector is related to its projection on itself. Now, in the case of these Euclidean vectors, we have a geometric understanding of what the "length" of a vector is, but for vectors of higher dimensions, that we cannot visualize, this other interpretation of the dot product is very useful. Through the dot product or in more general terms, the **inner product**, we can represent the length of any vector, even higher-dimensional ones, that we may not be able to visualize.

---

[2]Note here that we really distribute each term to perform the multiplication but something interesting occurs when we multiply components that have different unit vectors: Their dot product is 0 because they are orthogonal and $\cos 90° = 0$. So, the only terms of the components that have the *same* unit vector survive, since their dot product is 1 ($\cos 0° = 1$, and if it's the same unit vector, the angle difference between them is $0°$).

If we wanted to consider what percentage of one vector overlaps with the other, then we can simply *normalize* the two vectors we are working with[3]:

$$\frac{\vec{P}}{\sqrt{\vec{P} \cdot \vec{P}}} \cdot \frac{\vec{Q}}{\sqrt{\vec{Q} \cdot \vec{Q}}} = \cos \alpha$$

This quantity is between -1 and 1. The resulting value tells us the percentage of the overlap between the two vectors and the sign tells us whether the vectors are parallel or $0°$ apart (positive sign) or anti-parallel or $180°$ apart(negative sign). If the vectors are **orthogonal** or $90°$ apart, there is no overlap, so we would get 0. This is the utility of the dot product. It allows us to distinguish parallel and orthogonal vectors of any dimension very easily!

N-dimension Generalization

Now, let's stretch our understanding of vectors to higher dimensions. There's no reason why a vector should just be three-dimensional. Vectors are abstract objects in mathematics and can "live" in finite N-dimensional spaces or even in an infinite space. We can conceptually understand this as needing more of those orthogonal axes to describe our vector because they have more components. The downside is that if we add more dimensions, *we* cannot visualize this any longer, but the mathematics does not constrain itself to visualization, the same principle that applies to 3D then applies to N-D. N-dimensional vectors are very important in quantum computing, machine learning, physics, and more. For example, let's suppose we have oranges, bananas, apples, peaches, and strawberries and we want to keep track of how much of each fruit we have. We can use a 5D vector to do so because we have five kinds of fruits, and we can encode this information to a vector with five components, each corresponding to the fruits we have.

The last consideration is the fact that we can generalize vectors from real elements to complex ones. In the case of a 3D Cartesian

---

[3]Normalization just means you scale the vector by its magnitude so that its length equals 1.

coordinate system, the vector components x, y, and z are real numbers. But we may also have vectors that are defined more generally. Such vectors will have complex numbers of the form $a + bi$ as components. In this case, these vectors would naturally live in a *complex vector space*.

Checkpoint Exercises

1. Let's consider the directions, up and down represented by $\hat{N}$, a unit vector representing the north direction, right and left as $\hat{E}$, a unit vector representing east direction, and a unit vector representing forward and backward directions as $\hat{F}$. Suppose Sylvia moves 4 units east, 3 units backward, and 5 units north.

   a. Write down the vector $\vec{D}$, representing Sylvia's displacement and compute the magnitude of the vector.
   b. Now suppose we change the unit vectors to $\hat{NE}$, $\hat{NW}$, and $\hat{F}$, where $\hat{NE}$ and $\hat{NW}$ are the unit vectors that are 45° from the original $\hat{N}$ axis. Compute the magnitude of the vector. Comment on your results.

2. Suppose we have two vectors $\vec{P} = \hat{i} + 3\hat{j} + 4\hat{k}$ and $\vec{Q} = 6\hat{i} + 3\hat{j} + 2\hat{k}$. Compute the sum of the two vector, $\vec{P} + \vec{Q}$, and the magnitude of the resulting vector, $\vec{R}$.

3. What percentage of the two pairs of vectors overlaps? Are they orthogonal?

   a. $\vec{u} = \begin{pmatrix} 2 \\ 3 \\ 4 \end{pmatrix}$ and $\vec{v} = \begin{pmatrix} 2 \\ -6 \\ 7 \end{pmatrix}$.

   b. $\hat{i}$ and $\vec{p} = (0, 4, 1)$.

2.1.2   Dirac Notation

Quantum states are represented as vectors that live in a complex vector space known as the Hilbert space. The Hilbert space has useful mathematical properties that guide the physical interpretation of quantum states. For starters though, it is important to first become familiar with the notation that will follow throughout the rest of the book. This notation is known as *Dirac notation*. Paul Dirac, a famous British theoretical physicist, developed this notation to specifically label vectors that belong to the Hilbert space.

Bras and Kets

Using Dirac notation, a *bra* is a row vector labeled using the $\langle \,\, |$:

$$\langle u | = (\dots)$$

A column vector is known as an *ket* and is labeled using the $| \,\, \rangle$:

$$|v\rangle = \begin{pmatrix} : \end{pmatrix}$$

The components of $|v\rangle$, for example, may be a complex number, v_i, which has a corresponding complex conjugate v_i^*. If we write $|v\rangle$ and the corresponding $\langle v|$, we would have

$$|v\rangle = \begin{pmatrix} v_1 \\ v_2 \\ v_3 \\ \vdots \\ v_n \end{pmatrix} \qquad \langle v| = \begin{pmatrix} v_1^* & v_2^* & v_3^* & \cdots & v_n^* \end{pmatrix}$$

So, what we see here is that the *bra* of a *ket* vector is the **conjugate transpose** or **adjoint** of the *ket* vector. Transposing means that we turned the column into a row. Conjugation of a vector is taking the conjugate of each of the complex number components of the *ket* vector.[4]

An important operation that will often be performed is known as the *inner product*. The inner product is a more general version of the dot product for higher dimensional vector spaces. The dot product was a measure of overlap between two vectors. This means that for vectors that are $90°$ apart, the dot product would be zero because there is no overlap. However,

---

[4]Recall that the complex conjugate of a number $z = a + bi$ is $z^* = a - bi$ and is denoted by the * operator.

if the vectors are parallel, then the overlap is non-zero. In Eucledian vector spaces, we would define the dot product as follows[5]:

$$\vec{u} \cdot \vec{v} = |\vec{u}||\vec{v}| \cos \theta$$

If we have two ket vectors, $|u\rangle$ and $|v\rangle$, then their inner product is defined as follows:

$$\langle u|v \rangle = \begin{pmatrix} u_1^* & u_2^* & u_3^* & \cdots & u_n^* \end{pmatrix} \begin{pmatrix} v_1 \\ v_2 \\ v_3 \\ \vdots \\ v_n \end{pmatrix} = u_1^* v_1 + u_2^* v_2 + \cdots + u_n^* v_n$$

To compute the magnitude of such a complex vector, $|u\rangle$, we may in a similar fashion perform the inner product with itself, $\langle u|u \rangle$, and take the square root, just as we did before with vectors whose elements were real numbers. Thus,

$$||u\rangle| = \sqrt{\langle u|u \rangle} = \sqrt{u_1^* u_1 + u_2^* u_2 + \cdots + u_n^* u_n}$$
$$||u\rangle| = \sqrt{|u_1|^2 + |u_2|^2 + \cdots + |u_n|^2}$$

So, to find the magnitude of the complex vector, we need to sum up the squared magnitude of each of its complex components and take the square root. This is just like the distance formula for the vectors whose entries were real numbers. Thus, to make any complex vector, $|u\rangle$, a unit vector, $|n\rangle$, we normalize the vector:

$$|n\rangle = \frac{|u\rangle}{\sqrt{\langle u|u \rangle}}$$

---

[5]Note that the overlap would be determined by the cosine of the angle between the vectors because the cosine corresponds to the adjacent side and $\cos 90° = 0$, as we explained before.

Checkpoint Exercises

1. What is the *bra* vector for $|u\rangle = \begin{pmatrix} 1+i \\ 4 \end{pmatrix}$?

2. What is the *ket* vector for $\langle v| = \left((0.5 - i) \quad 0.2 \quad (1+0.5i) \quad 0.2 \right)$?

3. Suppose we have two complex vectors:

$$|P\rangle = \begin{pmatrix} 1 \\ 2+i \\ i \end{pmatrix} \qquad |Q\rangle = \frac{1}{2} \begin{pmatrix} 2+2i \\ 4 \\ 6+2i \end{pmatrix}$$

Normalize the vectors and then compute their inner product: $\langle P|Q\rangle$.

2.1.3   Superposition Principle

Superposition represents the sum of two or more physical quantities. This sum makes up a third quantity which is different from the original two. An intuitive picture of superposition is shown in Fig. 2.5 using classical waves.

As we can see in Fig. 2.5 when the peaks of the two waves align with each other, the resultant wave has a larger amplitude, indicating constructive interference. When the peaks of the waves do not align at all, but the waves have the same amplitude, then there is full cancellation which indicates destructive interference. These are the two extremes, but if one of the waves has some phase shift that causes

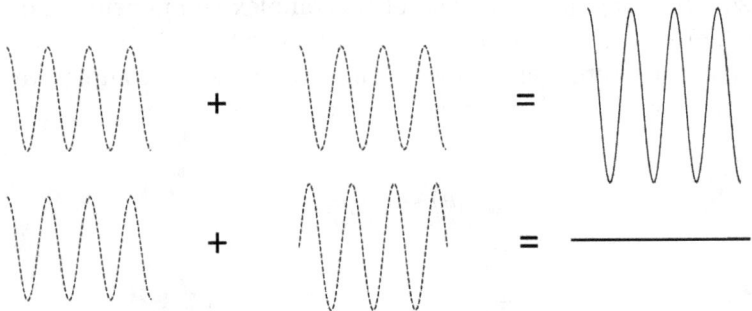

Figure 2.5.   In the first row, the amplitudes of the two waves add, implying *constructive interference*. In the second row, the amplitudes of the two waves cancel each other out, so we get *destructive interference*.

the amplitudes to add up in some places and cancel out in others, then we have a mixture of constructive and destructive interference.

We can also think about superposition in a more quantitative manner using vectors. Let's suppose we want to represent a column vector $|v\rangle = \begin{pmatrix} 3 \\ 4 \end{pmatrix}$, we can write this vector as

$$|v\rangle = 3 \begin{pmatrix} 1 \\ 0 \end{pmatrix} + 4 \begin{pmatrix} 0 \\ 1 \end{pmatrix}$$

where using the two vectors $\begin{pmatrix} 1 \\ 0 \end{pmatrix}, \begin{pmatrix} 0 \\ 1 \end{pmatrix}$, we may represent any arbitrary two-dimensional column vectors, $|\psi\rangle$, simply by changing the scaling factors in front:

$$|\psi\rangle = c_1 \begin{pmatrix} 1 \\ 0 \end{pmatrix} + c_2 \begin{pmatrix} 0 \\ 1 \end{pmatrix} = \begin{pmatrix} c_1 \\ c_2 \end{pmatrix}$$

If we allow these factors, c_1 and c_2, to be complex, then $\left\{ \begin{pmatrix} 1 \\ 0 \end{pmatrix}, \begin{pmatrix} 0 \\ 1 \end{pmatrix} \right\}$ are the spanning set for a two-dimensional complex vector space which we can denote $\mathbb{C}^2$. These vectors are **linearly independent**, which means that there is no way to relate the two vectors to each other through multiplying them by some scalar. In fact, these two vectors also happen to be orthogonal, which you may check by taking their inner product. In fact, this set of vectors also forms a **basis**, which is the minimum number of linearly independent vectors needed to represent any other vector (with the same number of dimensions). This is the same idea as when we represented real vectors in terms of the $\hat{i}$, $\hat{j}$, and $\hat{k}$ unit vectors. These three vectors formed a basis for a real 3D vector space.

Quantum objects are represented as vectors which means that they obey the superposition principle and all concepts we have discussed thus far. In particular, a 2D quantum state $|\psi\rangle$ can be written as a superposition of the basis vectors $\left\{ \begin{pmatrix} 1 \\ 0 \end{pmatrix}, \begin{pmatrix} 0 \\ 1 \end{pmatrix} \right\}$ or as we later call them $\{|0\rangle, |1\rangle\}$. One reason this basis is used is because the two vectors are orthogonal to each other (they do not overlap) so the component in one direction does not affect in one direction does not affect the other direction.

The quantum state must also be a unit vector because of the probabilistic interpretation that we will learn in detail in the next chapter. Basically, we cannot consider probabilities that are greater than 1. Thus, the components that scale the basis vectors, c_1 and c_2, must satisfy the normalization condition that makes $|\psi\rangle$ a unit vector:

$$|c_1|^2 + |c_2|^2 = 1$$

Again, note that we must take the magnitude of the scaling factors because they may be complex numbers!

Checkpoint Exercises

1. Determine if the following sets of vectors are linearly independent:

   a. $|v_1\rangle = \begin{pmatrix} 2 \\ 1 \end{pmatrix}$ $|v_2\rangle = \begin{pmatrix} 6 \\ 3 \end{pmatrix}$.

   b. $|v_1\rangle = \begin{pmatrix} 5 \\ 2 \end{pmatrix}$ $|v_2\rangle = \begin{pmatrix} 10 \\ 2 \end{pmatrix}$.

2. Suppose we want to represent a vector $|\psi\rangle = \begin{pmatrix} 15 \\ 2 \end{pmatrix}$ in terms of the two linearly independent vectors identified in Question 1. What are the scaling factors c_1 and c_2 for the vectors $|v_1\rangle$ and $|v_2\rangle$ that can be superposed to give us $|\psi\rangle$?

3. Suppose we have a quantum state that is not normalized $|\psi\rangle = \begin{pmatrix} i \\ 1 \end{pmatrix}$. Write the state in the $\left\{ \begin{pmatrix} 1 \\ 1 \end{pmatrix}, \begin{pmatrix} 1 \\ -1 \end{pmatrix} \right\}$ basis and find c_1 and c_2 that normalize the state $|\psi\rangle$.

4. How is the orthogonality of two orthonormal quantum states, $|u\rangle$ and $|v\rangle$, defined? Use the following function in your answer, which is called the Kronecker delta function:

$$\delta_{ij} = \begin{cases} 0, & i \neq j \\ 1, & i = j \end{cases}$$

2.2 Single Qubit Control with Matrices

Classical devices and computers work with the fundamental unit called the **bit**. Physically, bits are digital signals that transistors (which are voltage-controllable switches) toggle back and forth. Bits can either have a value of 0 or 1 (representing ON or OFF states of the switch), and in this sense, they represent classical **two-state** or **two-level** systems. Through combinations of transistors, resistors, and diodes, digital logic gates are constructed to manipulate data and perform basic computation like addition or multiplication.

In quantum computing, we deal with quantum bits or *qubits*. While the bit represents a classical two-level system, a qubit represents a quantum two-level system. A formal definition of a quantum two-level system is two quantized energy levels which obey Schrödinger's equation. In most cases, quantum mechanical properties appear at small scales (nanometer size). These are the scales where the wave–particle duality shows up. Interestingly enough, new transistors are also reaching these scales and quantum effects should even be considered in current chips.

We discussed that quantum objects can exhibit both particle-like and wave-like properties, so they can be represented by $|0\rangle$, $|1\rangle$, or a superposition of $|0\rangle$ and $|1\rangle$. This superposition is the *state space* of the qubit which describes the set of all possible states the qubit could be in. We start off explaining this concept using the coin toss analogy.

We can think of the states $|0\rangle$ and $|1\rangle$ to be heads and tails, respectively. These are the same as the bit states 0 and 1, as shown in Fig. 2.6.

In the case of a qubit, we can have superposition states which are effectively a continuum of the different possible orientations a coin

Figure 2.6. Head represents the 0 bit state or qubit state $|0\rangle$ and tail represents the 1 bit state or qubit state $|1\rangle$.

Figure 2.7.   Continuum of different coin orientations when it is tossed.

can take when it is tossed. The qubit state could be anywhere on this spectrum, as represented in Fig. 2.7.

The superposition is a sum of $|0\rangle$ and $|1\rangle$ with scaling factors that indicate how much of state $|0\rangle$ and how much of state $|1\rangle$ makes up a general qubit state, $|\psi\rangle$. To continue with the coin toss analogy, we are thinking about the state of the qubit as some arbitrary orientation of the coin and somehow we are keeping track of how close (in probabilistic terms) it is to be heads vs. tails for one particular trial.

2.3   Basis Vectors

The general single qubit state is represented as a complex vector. The physical qubit itself may be an electron spin or the photon polarization states, but both of these examples are mathematically represented as vectors. A useful way to define a general qubit state is to first specify a convenient basis to work in. For a single qubit, we need two basis vectors to specify this two-dimensional vector space that the qubit exists in. This space is complex because we can scale our basis vectors with complex numbers. We take a superposition of the basis vectors to generate arbitrary qubit states.

Understanding bases is a critical part of the content moving forward. Intuitively, the basis can be thought of as the minimum set of vectors that defines the space. For example, for real 3D vectors, we had 3 unit vectors that we scaled and summed to specify any other 3D vector. So, in this case, the basis was $\{\hat{i}, \hat{j}, \hat{k}\}$. With a single qubit, the most convenient basis to work with is the $\{|0\rangle, |1\rangle\}$ basis.

$$|0\rangle = \begin{pmatrix} 1 \\ 0 \end{pmatrix}$$

$$|1\rangle = \begin{pmatrix} 0 \\ 1 \end{pmatrix}$$

Now, any other qubit state can be formed as a linear combination or *superposition* of the basis states $|0\rangle$ and $|1\rangle$. The general qubit state, $|\psi\rangle$, can be written as column vectors as follows:

$$|\psi\rangle = c_1 |0\rangle + c_1 |1\rangle = \begin{pmatrix} c_1 \\ c_2 \end{pmatrix}$$

Any $|\psi\rangle$ that represents a quantum state must be a unit vector and therefore its magnitude (or squared magnitude) must be 1:

$$|\,|\psi\rangle\,|^2 = 1$$

We showed in the previous section that this constraint in the magnitude of quantum states gives rise to the **normalization condition**, where the squared magnitudes of the scaling factors must add up to 1:

$$|c_1|^2 + |c_2|^2 = 1$$

The scaling factors in front of the basis vectors, c_1 and c_2, are known as **probability amplitudes** and they may be complex numbers. This is why we need to take their magnitudes. The value $|c_1|^2$ gives the probability of getting $|0\rangle$ when we measure $|\psi\rangle$ and $|c_2|^2$ gives us the probability of getting $|1\rangle$ when we measure $|\psi\rangle$.

Now, in order to do anything useful with qubits, we have to be able to control them or perform some operation to change its state from one to the other. But how can we perform operations on vectors? Well, we need matrices!

2.4   Fundamentals of Matrices

A **matrix** is a rectangular array of numbers or symbols that consists of rows and columns. One can write the most general matrix as follows, where there are $1, 2, \ldots, m$ rows and $1, 2, \ldots, n$ columns:

$$\mathbf{A} = \begin{pmatrix} a_{11} & a_{12} & \cdots & a_{1n} \\ a_{21} & a_{22} & \cdots & a_{2n} \\ \vdots & \vdots & \ddots & \vdots \\ a_{m1} & a_{m2} & \cdots & a_{mn} \end{pmatrix}$$

The number of rows and columns defines the **dimensions** of the matrix. Matrices are generally labeled by a bold-faced, uppercase letter, while the matrix elements are denoted by a lowercase letter in subscripts. The subscripts label the location of the elements within the matrix. For example, a_{23} would be an element in the second row and third column (right after a_{22}). When using these matrices on qubits, we will deal with **square matrices**, which are matrices that have the same number of rows as they do columns, thus $m = n$.

2.4.1   Working with Matrices

Let's go through some of the mechanics of working with matrices [24].

Notation

Suppose we have three matrices of the same dimension: **A**, **B**, and **C**.

1. $\mathbf{A}^\mathsf{T}$ represents the transpose of matrix **A**. This action flips a matrix over its diagonal, for example,

$$\mathbf{A} = \begin{bmatrix} 1 & 4 \\ 6 & 8 \end{bmatrix} \quad \mathbf{A}^\mathsf{T} = \begin{bmatrix} 1 & 6 \\ 4 & 8 \end{bmatrix}$$

2. $\mathbb{I}$ represents the identity matrix which is the matrix that is equivalent of the number 1:

$$\mathbb{I} = \begin{bmatrix} 1 & 0 & \cdots & 0 \\ 0 & 1 & \cdots & 0 \\ \vdots & \vdots & \ddots & \vdots \\ 0 & 0 & \cdots & 1 \end{bmatrix}$$

3. $\mathbf{A}^{-1}$ represents the inverse of a matrix. The matrix inverse exists only for *square* matrices with non-zero determinants.[6] The matrix inverse has the property $\mathbf{A}^{-1}\mathbf{A} = \mathbb{I} = \mathbf{A}\mathbf{A}^{-1}$.

Matrix Addition

Suppose we wanted to add the following two matrices, $\mathbf{A}$ and $\mathbf{B}$:

$$\mathbf{A} + \mathbf{B} = \begin{bmatrix} 1 & 4 \\ 6 & 8 \end{bmatrix} + \begin{bmatrix} 1 & 5 \\ 4 & 8 \end{bmatrix} = \begin{bmatrix} 2 & 9 \\ 10 & 16 \end{bmatrix}$$

As you can see, matrix addition works by adding the elements that are in the same corresponding rows and columns. Thus, addition is commutative, meaning that $\mathbf{A} + \mathbf{B} = \mathbf{B} + \mathbf{A}$. Note, however, that if the dimensions of the matrix were not the same, then we would not be able to add them which is the same as with vector addition. Matrix addition is also associative, so if we had another 2×2 matrix, $\mathbf{C}$, and added it to $\mathbf{A}$ and $\mathbf{B}$, we would find that

$$\mathbf{A} + \mathbf{B} + \mathbf{C} = \mathbf{A} + (\mathbf{B} + \mathbf{C}) = (\mathbf{A} + \mathbf{B}) + \mathbf{C}$$

Matrix Multiplication

This is the most important computation we will be doing with matrices. To multiply two matrices, first we take the row of the first matrix

---

[6]The **determinant** is a scalar quantity that tells us how the space is being stretched or compressed. For example, the determinant of a 2×2 matrix represents the area of one grid of the space. It is area because we are in 2D. For 3×3 matrix, the determinant corresponds to the volume and so on.

and perform a dot product with the column of the second matrix and so on. Here is a concrete example:

$$\mathbf{AB} = \begin{bmatrix} 1 & 4 \\ 6 & 8 \end{bmatrix} \begin{bmatrix} 1 & 5 \\ 4 & 8 \end{bmatrix} = \begin{bmatrix} (1*1)+(4*4) & (1*5)+(4*8) \\ (6*1)+(8*4) & (6*5)+(8*8) \end{bmatrix} = \begin{bmatrix} 17 & 37 \\ 38 & 94 \end{bmatrix}$$

What if we took $\mathbf{BA}$?

$$\mathbf{BA} = \begin{bmatrix} 1 & 5 \\ 4 & 8 \end{bmatrix} \begin{bmatrix} 1 & 4 \\ 6 & 8 \end{bmatrix} = \begin{bmatrix} (1*1)+(5*6) & (1*4)+(5*8) \\ (4*1)+(8*6) & (4*4)+(8*8) \end{bmatrix} = \begin{bmatrix} 31 & 44 \\ 52 & 80 \end{bmatrix}$$

As we can see, multiplication is NOT commutative for matrices. The multiplication cares about the order.

The dimensions of the matrices are also important when doing matrix multiplication. Note that the number of columns in the first matrix must match the number of rows in the second matrix in order to multiply the two matrices. For example, a 2×2 matrix can be multiplied to any matrix with two rows. So, in general, if we have two matrices with dimensions $m_1 \times n_1$ and $m_2 \times n_2$, then in order for the multiplication to work out, the number of columns of the first matrix, n_1, must be equal to the number of rows in the second matrix, m_2. The dimensions of the final answer will be $m_1 \times n_2$. Lastly, if the matrix has a scalar in front, each element of the matrix will be scaled by that factor just as with vectors.

Continuing forward, we are concerned specifically with square matrices and will introduce special types of square matrices that are used to control and change the state of qubits.

Checkpoint Exercises

Perform the indicated operations on the following matrices if possible:

$$\mathbf{A} = \begin{pmatrix} -8 & -6 \\ 7 & 3 \end{pmatrix} \quad \mathbf{B} = \begin{pmatrix} 9 & -1 \\ 5 & 0 \end{pmatrix} \quad \mathbf{C} = \begin{pmatrix} 2 & -2 \\ 4 & 1 \end{pmatrix} \quad \mathbf{D} = \begin{pmatrix} 2 & -2 \\ 4 & 1 \\ 3 & 0 \end{pmatrix}$$

1. $\mathbf{A}^\mathsf{T} + \mathbf{B} + \mathbf{C}$,
2. $\mathbf{B} + \mathbf{D}$,
3. $\mathbf{AD}^\mathsf{T}$,
4. $\mathbf{DA}$,
5. $3\mathbf{C}$.

2.4.2   Linear Transformations

As we saw, qubits are represented by complex vectors and the space in which they live in is a *complex vector space* known as the Hilbert space. For a single qubit, the space is called $\mathbb{C}^2$ because we need two complex numbers which scale the basis vectors to specify an arbitrary single qubit state, $|\psi\rangle$. We can use matrices to change the state of qubits. A single qubit is represented as a 2×1 column vector, so we need a 2×2 matrix, M (since we want the final state to be of the same dimension as the one we started with) to take us from the state $|\psi\rangle$ to $|\psi'\rangle$:

$$|\psi'\rangle = M |\psi\rangle$$

Matrices could be regarded of as *linear transformations* which in the world of quantum mechanics are known as **operators**. They are used to manipulate quantum states, such as our qubits. The nature of quantum systems requires all these matrices to be reversible. We discuss the important property of linearity, learn to calculate special values and vectors for operators known as eigenvalues and eigenvectors, and discuss the properties of unitary and Hermitian matrices.

Understanding Linearity

Linearity itself is an intuitive concept. If we have some transformation or function that takes in certain inputs, and these inputs are scaled by some factor or two or more inputs are added together, the output of the function *responds* in the same way! One key feature of a linear transformation is that if the input is zero, then the output must also be zero. If nothing is happening to the input, then nothing should happen to the output.

Suppose we have a transformation, T, taking a sum of input vectors $\vec{x}_1 + \vec{x}_2$, if the function is linear, then the result will be the sum of each output plugged into the function separately:

$$T(\vec{x}_1 + \vec{x}_2) = T(\vec{x}_1) + T(\vec{x}_2)$$

Also, let's say we wanted to know what happened if the input vector, $\vec{x}_1$, was scaled by some factor, c, then

$$T(c\vec{x}_1) = cT(\vec{x}_1)$$

This is what is meant by "the function *responds* in the same way." If the input is a superposition of input vectors, then the result is the superposition of the output vectors. In a more general sense, linearity is defined as

$$T(c\vec{x}_1 + d\vec{x}_2) = cT(\vec{x}_1) + dT(\vec{x}_2)$$

So, for a transformation to be linear, it must satisfy the following[7]:

$$T(\vec{0}) = \vec{0}$$

$$T(c\vec{x}_1 + d\vec{x}_2) = cT(\vec{x}_1) + dT(\vec{x}_2)$$

A nonlinear transformation would be a transformation that does not satisfy the properties above, for example, if we had a function $f(x) = mx + b$. Such a function would not satisfy the first condition for linearity because $f(0) = b$ and not zero. Functions such as $p(x) = x^2$ are also considered nonlinear because they do not satisfy the second property of linearity:

$$p(cx_1 + dx_2) = (cx_1 + dx_2)^2 \neq cp(x_1) + dp(x_2)$$

Matrix Transformations

When using matrices to transform vectors, the transformations will always be linear. In quantum mechanics, we will be concerned with transformations involving square matrices, and matrices that preserve the magnitude of a quantum state, because we always need the state to be normalized.

Suppose that we have the state $|0\rangle$ and we operate on it with the matrix

$$\mathbf{X} = \begin{pmatrix} 0 & 1 \\ 1 & 0 \end{pmatrix}$$

So, we would have the following:

$$\mathbf{X}|0\rangle = \begin{pmatrix} 0 & 1 \\ 1 & 0 \end{pmatrix} \begin{pmatrix} 1 \\ 0 \end{pmatrix} = \begin{pmatrix} 0 \\ 1 \end{pmatrix} = |1\rangle$$

As we can see, this matrix transformation changed the original vector, but the resulting vector still has a magnitude of 1 and the same

---

[7]$\vec{0}$ is called the **zero vector** and it is the vector whose elements are all zeros.

dimensions as the input vector. This matrix flipped the qubit from $|0\rangle$ to $|1\rangle$, so it essentially performed a rotation on the qubit. In fact, these kinds of matrices that keep the output vector in the same dimension and magnitude as the input vector can be thought of as *rotation matrices*.

What if we tried to operate on the state $|v\rangle = \frac{1}{\sqrt{2}} \begin{pmatrix} 1 \\ -1 \end{pmatrix}$:

$$\mathbf{X}\,|v\rangle = \begin{pmatrix} 0 & 1 \\ 1 & 0 \end{pmatrix} \left(\frac{1}{\sqrt{2}} \begin{pmatrix} 1 \\ -1 \end{pmatrix} \right) = \frac{1}{\sqrt{2}} \begin{pmatrix} 0 & 1 \\ 1 & 0 \end{pmatrix} \begin{pmatrix} 1 \\ -1 \end{pmatrix}$$

$$= \frac{1}{\sqrt{2}} \begin{pmatrix} -1 \\ 1 \end{pmatrix} = -\,|v\rangle$$

So, in this example, when we tried to operate on $|v\rangle$ with $\mathbf{X}$, we got the same vector back but scaled by a factor of -1. This means for this vector in particular, the matrix $\mathbf{X}$ only scales the vector. We learn in the following section that such a vector is known as an **eigenvector** and the value that it is scaled by is called an **eigenvalue** of $\mathbf{X}$.

Checkpoint Exercises

1. Operate on the vector $|u\rangle = \frac{1}{\sqrt{2}} \begin{pmatrix} 1 \\ i \end{pmatrix}$ with the matrix $\mathbf{X}$. Is $|u\rangle$ an eigenvector of $\mathbf{X}$?

2. Operate on the vector $|u\rangle = \frac{1}{\sqrt{2}} \begin{pmatrix} 1 \\ i \end{pmatrix}$ with the matrix $\mathbf{Y} = \begin{pmatrix} 0 & -i \\ i & 0 \end{pmatrix}$. Is $|u\rangle$ an eigenvector of $\mathbf{Y}$?

3. Operate on the vector $|\psi\rangle = c_1\,|0\rangle + c_2\,|1\rangle$ with the matrix $\mathbf{Z} = \begin{pmatrix} 1 & 0 \\ 0 & -1 \end{pmatrix}$. Is $|\psi\rangle$ an eigenvector of $\mathbf{Z}$?

2.4.3 Eigenvalues and Eigenvectors

In mathematics, the term *eigen* means "proper" or "characteristics." So, an eigenvector is a characteristic vector of some matrix or operator, $\mathbf{M}$, whereas an eigenvalue is a characteristic value of an operator, $\mathbf{M}$. For the 2×2 matrices, there are two eigenvalues and two corresponding eigenvectors. The relationship that represents this idea is

$$\mathbf{M} \left| u \right\rangle = \lambda \left| u \right\rangle$$

This means that there are some vectors $\left| u \right\rangle$ that when operated on by the matrix, $\mathbf{M}$, the result is the same vector *scaled* by some number, λ. $\left| u \right\rangle$ are the **eigenvectors** of $\mathbf{M}$ and the numbers λ are the **eigenvalues** of $\mathbf{M}$. Let's do an example to illustrate how we may calculate the eigenvalues and eigenvectors.

Suppose we have a matrix, $\mathbf{M} = \begin{bmatrix} 1 & 1 \\ 4 & 1 \end{bmatrix}$, and we want to find its eigenvectors, $\left| u_1 \right\rangle$ and $\left| u_2 \right\rangle$, and their corresponding eigenvalues, λ_1 and λ_2. The easiest way to think about the problem is as follows:

$$\begin{bmatrix} 1 & 1 \\ 4 & 1 \end{bmatrix} \begin{bmatrix} a \\ b \end{bmatrix} = \lambda \begin{bmatrix} a \\ b \end{bmatrix}$$

If we perform matrix multiplication on the left side and scalar multiplication on the right side, then we end up with the following linear system of equations:

$$a + b = \lambda a$$
$$4a + b = \lambda b$$

Solving the first equation for b, we have $b = a(\lambda - 1)$. We plug this into the second equation and simplify:

$$4a + a(\lambda - 1) = \lambda a(\lambda - 1)$$
$$3a = a(\lambda^2 - 2\lambda)$$

Cancelling out the a's, we have

$$\lambda^2 - 2\lambda - 3 = (\lambda - 3)(\lambda + 1) = 0$$

We have just solved for our eigenvalues. We can label $\lambda_1 = 3$ and $\lambda_2 = -1$.

Now, we can go back to the original matrix equation and solve the eigenvectors corresponding to these eigenvalues:

$$\begin{bmatrix} 1 & 1 \\ 4 & 1 \end{bmatrix} \begin{bmatrix} a \\ b \end{bmatrix} = 3 \begin{bmatrix} a \\ b \end{bmatrix} \rightarrow \begin{matrix} a + b = 3a \\ 4a + b = 3b \end{matrix}$$

$$\begin{bmatrix} 1 & 1 \\ 4 & 1 \end{bmatrix} \begin{bmatrix} a \\ b \end{bmatrix} = -1 \begin{bmatrix} a \\ b \end{bmatrix} \rightarrow \begin{matrix} a + b = -a \\ 4a + b = -b \end{matrix}$$

If we solve each set of linear equations, we find that the eigenvector $|u_1\rangle$ corresponding to λ_1 is $|u_1\rangle = \begin{bmatrix} 1 \\ 2 \end{bmatrix}$ and eigenvector $|u_2\rangle$ corresponding to λ_2 is $|u_2\rangle = \begin{bmatrix} 1 \\ -2 \end{bmatrix}$. We can easily check that these are correct by performing the matrix multiplication with each vector and ensuring that we get the same vector scaled by its corresponding eigenvalue.

The calculations of these values and vectors are straightforward, but what does this mean? Well, a matrix is *transforming* the space. Let's say we have our familiar 2D Euclidean coordinate plane that is defined by the unit vectors $\vec{e_1} = \begin{bmatrix} 1 \\ 0 \end{bmatrix}$ and $\vec{e_1} = \begin{bmatrix} 0 \\ 1 \end{bmatrix}$. If we transformed these vectors using M, then we would have

$$M\vec{e_1} = \begin{bmatrix} 1 \\ 4 \end{bmatrix}$$

$$M\vec{e_2} = \begin{bmatrix} 1 \\ 1 \end{bmatrix}$$

Basically, instead of having each grid in the plane have area of 1 as we would with the unit vectors, we would now have a new grid that is stretched and rotated. The area that it covers is given by the determinant of the matrix $\mathbf{M}$, as shown in Fig. 2.8.

The eigenvectors $|u_1\rangle$ and $|u_2\rangle$ have a special property where they only get stretched or compressed. In other words, they are simply scaled by the factor the eigenvalues λ_1 and λ_2 that correspond to each vector. They are not rotated or translated in any way. So, if we had a 2D plane defined that was using $|u_1\rangle$ and $|u_2\rangle$, and we applied $\mathbf{M}$, nothing would change about the orientation of the space.

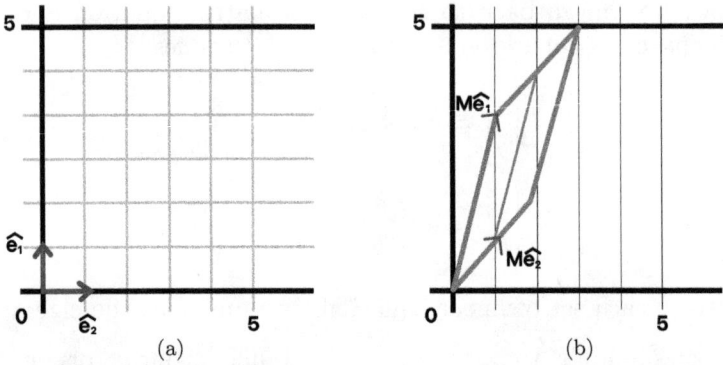

Figure 2.8.　(a) The unit vectors, $\hat{e}_1$ and $\hat{e}_2$, form our familiar 2D coordinate grid, where each grid has an area of 1; (b) the matrix changes this grid and now the new area of one grid is the determinant of the matrix $\mathbf{M}$.

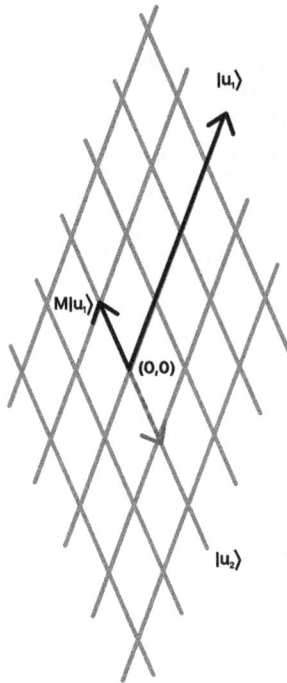

Figure 2.9.　Grid is defined in terms of the eigenvectors, $|u_1\rangle$ and $|u_2\rangle$. When matrix $\mathbf{M}$ acts on these vectors (shown in black), it only scales them but does not change the orientation of the grid.

When we define the space in terms of the eigenvectors, this is called the **eigenspace**. The visual representation of this is given in Fig. 2.9.

Characteristic Equation

A more ubiquitous way to calculate the eigenvalues of a matrix is to use the characteristic polynomial. We start again with our eigenvalue equation:

$$\mathbf{M}\,|u\rangle = \lambda\,|u\rangle$$

We can move the $\lambda\,|u\rangle$ term to the left and factor our $|u\rangle$[8]:

$$(\mathbf{M} - \lambda\mathbb{I})\,|u\rangle = 0$$

Now, vector $|u\rangle$ cannot be the **0** vector if it is to be an eigenvector of matrix $\mathbf{M}$. However, the equation implies that when this matrix $(\mathbf{M} - \lambda\mathbb{I})$ operates on $|u\rangle$, we get the **0** vector, which means that $(\mathbf{M} - \lambda\mathbb{I})$ must not have an inverse.[9] A matrix without an inverse has a determinant of 0. Thus, we have our characteristic equation:

$$\det(\mathbf{M} - \lambda\mathbb{I}) = 0$$

The determinant of a general 2×2 matrix, $\mathbf{A}$, can be computed as follows:

$$\det(\mathbf{A}) = \begin{vmatrix} a & b \\ c & d \end{vmatrix} = ad - bc$$

Let's look at another example of finding the eigenvalues but directly using the characteristic equation method.

Suppose we have the matrix $\mathbf{D} = \begin{pmatrix} 1 & 0 \\ 1 & -1 \end{pmatrix}$ and we want to find its eigenvalues and eigenvectors. We would have the following eigenvalue

---

[8]Note here that we are multiplying λ by I which is the identity matrix that we mentioned before as being like multiplying by the number 1.

[9]For a matrix inverse to exist, we must have that $A\,|v\rangle = \mathbf{0}$, where $|v\rangle = \mathbf{0}$. Otherwise, the inverse of matrix A does not exist.

equation:

$$\mathbf{D}\,|v\rangle = \lambda\,|v\rangle$$

Rearranging the equation by moving the $\lambda\,|v\rangle$ term to the left and factoring out $|v\rangle$,

$$(\mathbf{D} - \lambda\mathbb{I})\,|v\rangle = 0$$

Now, we just need to find the determinant of the matrix:

$$(\mathbf{D} - \lambda\mathbb{I}) = \begin{pmatrix} 1 & 0 \\ 1 & -1 \end{pmatrix} - \begin{pmatrix} \lambda & 0 \\ 0 & \lambda \end{pmatrix} = \begin{pmatrix} 1-\lambda & 0 \\ 1 & -1-\lambda \end{pmatrix}$$

$$\det(\mathbf{D} - \lambda\mathbb{I}) = \begin{vmatrix} 1-\lambda & 0 \\ 1 & -1-\lambda \end{vmatrix} = (1-\lambda)(-1-\lambda) + 0$$

$$= -1 - \lambda + \lambda + \lambda^2$$

$$= \lambda^2 - 1 = 0$$

Thus, $\lambda = \pm 1$.

Checkpoint Exercises

1. Find the eigenvectors for matrix $\mathbf{D}$.
2. Find the eigenvalues and eigenvectors of the matrix $\mathbf{P} = \begin{pmatrix} 1 & 1 \\ 1 & -1 \end{pmatrix}$.

2.4.4 Properties of Unitary Matrices

A unitary matrix , U, is a special kind of square matrix that is used throughout quantum computing. It has the following property:

$$\mathbf{U}^\dagger\mathbf{U} = \mathbf{U}\mathbf{U}^\dagger = \mathbb{I}$$

The $\dagger$ symbol denotes conjugate transpose operation (also called *Hermitian adjoint*) and is denoted by

$$\dagger \longleftrightarrow *^\mathsf{T}$$

As it's defined, first, we take the conjugate of each of the elements of the matrix[10] and then we transpose the matrix.

---

[10]Elements may be complex numbers, so $(a+bi)^* = (a-bi)$. If they are real numbers, then $(a)^* = a$.

Let's consider the matrix $\mathbf{U} = \frac{1}{\sqrt{2}} \begin{pmatrix} 1 & 1 \\ i & -i \end{pmatrix}$ and check if it is unitary. We need to find the adjoint of U, so we first take the conjugate of the matrix and then we transpose:

$$\mathbf{U}^\dagger = \frac{1}{\sqrt{2}} \left[\left(\begin{pmatrix} 1 & 1 \\ i & -i \end{pmatrix} \right)^* \right]^\mathsf{T} = \frac{1}{\sqrt{2}} \left[\begin{pmatrix} 1 & 1 \\ -i & i \end{pmatrix} \right]^\mathsf{T} = \frac{1}{\sqrt{2}} \begin{pmatrix} 1 & -i \\ 1 & i \end{pmatrix}$$

Now, we can verify:

$$\mathbf{U}^\dagger \mathbf{U} = \left(\frac{1}{\sqrt{2}} \cdot \frac{1}{\sqrt{2}} \right) \begin{pmatrix} 1 & -i \\ 1 & i \end{pmatrix} \begin{pmatrix} 1 & 1 \\ i & -i \end{pmatrix}$$

$$= \frac{1}{2} \begin{pmatrix} 1^2 - i^2 & 1^2 + i^2 \\ 1^2 + i^2 & 1^2 - i^2 \end{pmatrix}$$

$$= \frac{1}{2} \begin{pmatrix} 2 & 0 \\ 0 & 2 \end{pmatrix} = \begin{pmatrix} 1 & 0 \\ 0 & 1 \end{pmatrix} = \mathbb{I}$$

We have shown that the above matrix $\mathbf{U}$ is unitary. Recall from the complex numbers section that we interpreted multiplication by complex numbers as rotation. In fact, the reason why unitary matrices are so useful is because they perform rotations around a unit circle while preserving the magnitude of the vector.

Hermitian Matrices

There is another special class of matrices used in quantum computing known as **Hermitian matrices**. These matrices or operators are very important in quantum mechanics because they represent *observables*. Observables are Hermitian operators that correspond to some physical quantity of the system, such as spin, energy, position, or momentum.

Hermitian operators have the property that they are equal to their conjugate transpose. So, a Hermitian matrix $\mathbf{H} = \mathbf{H}^\dagger$. This is true if the diagonal elements of the matrix are real numbers and the off-diagonal elements are complex conjugates of each other. Remember that when we transpose a matrix, the diagonal elements do not change, and if we want the original matrix to be equal to its complex conjugate, this means that the diagonal elements have to be real so that $a^* = a$. The off-diagonal elements have to be complex conjugates of each other so that if we apply the conjugate operation and then transpose, we end up with the same matrix:

$$\mathbf{H}^\dagger = \frac{1}{\sqrt{2}} \left[\left(\begin{bmatrix} 1 & 1+i \\ 1-i & -1 \end{bmatrix} \right)^* \right]^\mathsf{T} = \frac{1}{\sqrt{2}} \begin{bmatrix} 1 & 1-i \\ 1+i & -1 \end{bmatrix}^\mathsf{T}$$

$$= \frac{1}{\sqrt{2}} \begin{bmatrix} 1 & 1+i \\ 1-i & -1 \end{bmatrix} = \mathbf{H}$$

Since Hermitian matrices are equal to their conjugate transpose, one can prove that they possess only *real* eigenvalues. Let's see this by finding the eigenvalues of matrix $\mathbf{H}$ using what we learned in the previous section:

$$\det(\mathbf{H} - \lambda\mathbb{I}) = \begin{vmatrix} \left(\frac{1}{\sqrt{2}} - \lambda\right) & \frac{1}{\sqrt{2}}(1+i) \\ \frac{1}{\sqrt{2}}(1-i) & \left(-\frac{1}{\sqrt{2}} - \lambda\right) \end{vmatrix} = -\left(\frac{1}{\sqrt{2}} - \lambda\right)\left(\frac{1}{\sqrt{2}} + \lambda\right)$$

$$-\frac{1}{2}(1+i)(1-i) = 0$$

Simplifying the characteristic polynomial further, we have[11]

$$\lambda^2 - \frac{3}{2} = 0$$

Therefore, we obtain the following eigenvalues for matrix $\mathbf{H}$ which are real:

$$\lambda_1 = \sqrt{\frac{3}{2}} \qquad \lambda_2 = -\sqrt{\frac{3}{2}}$$

Checkpoint Exercises

Determine whether the following matrices are unitary or Hermitian and find their eigenvalues using the characteristic equation and then find the corresponding eigenvectors:

---

[11]Note that when multiplying a complex number by its conjugate,

$$(a + ib)(a - ib) = a^2 - iab + iab - (ib)^2 = a^2 + -i^2 b^2 = a^2 + b^2$$

1. $\mathbf{Y} = \begin{pmatrix} 0 & -i \\ i & 0 \end{pmatrix}$.

2. $\mathbf{F} = \begin{pmatrix} 0 & 1 \\ i & 0 \end{pmatrix}$.

Homework 2

1. The vertices of a triangle are give by the points $A = (3, 1, 1)$; $B = (4, 6, -1)$; $C = (2, 3, 3)$. Find the three interior angles of the angles. (*Hint*: Assign vectors to each side of the triangle by subtracting the points. Then, use the dot product in the two different ways we learned to solve for the angle.)

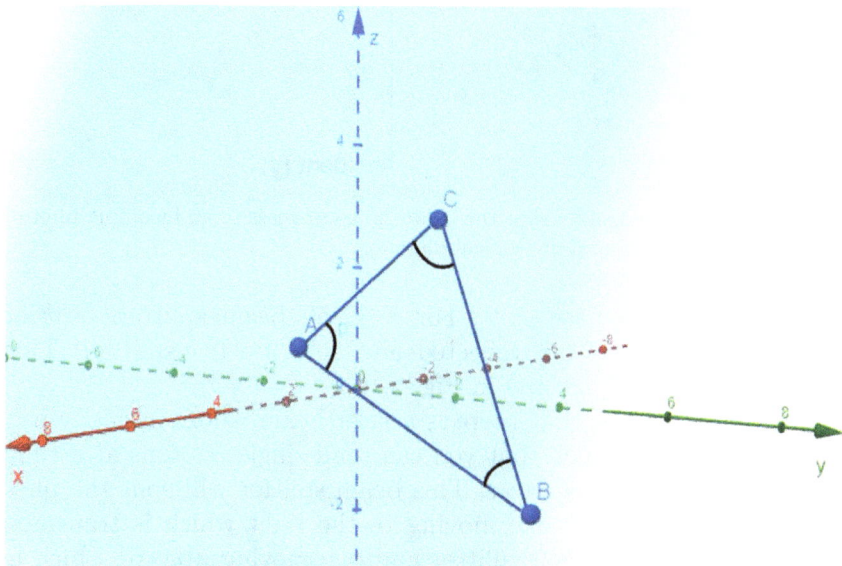

2. Suppose we have a symmetric beam splitter[12] on each side that the photon is reflected, it would pick up a $90° = \frac{\pi}{2}$ rad $= 1.57$ rad phase shift which can be represented with the complex number, $e^{i1.57} = i$. When the photon is transmitted, the beam splitter

---

[12]The beam splitter is an important optical component used to split light into different directions and recombine light coming in from two different directions.

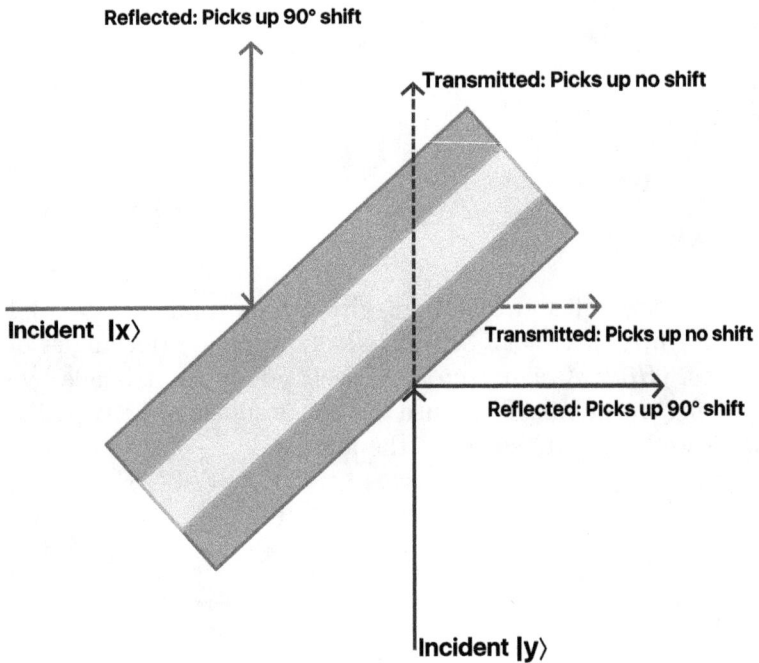

Figure 2.10. Phase shifts that the beam splitter causes to an incoming photon polarized in the horizontal or vertical direction.

introduced no phase shift. For a 50/50 beam splitter, 50% of incoming photons are reflected and 50% are transmitted. This beam splitter is pictured in Fig. 2.10.

Let us consider that there is a laser beam whose intensity can be turned down such that you can send single photons at a time through the beam splitter. This beam splitter will split the photon into two paths: one moving to the right which is transmitted through the beam splitter and one moving upward which is reflected off the beam splitter. Let $|x\rangle$ be the state of the photon moving to the right (horizontal) and $|y\rangle$ be the state of the photon moving upward (vertical). If $|x\rangle$ goes through the beam splitter, the resulting state is an equal superposition between the horizontal and vertical states with a 90° phase shift on the vertical state, since it is reflected:

$$|P0\rangle = \frac{1}{\sqrt{2}}(|x\rangle + i\,|y\rangle)$$

If $|y\rangle$ goes through the beam splitter, the resulting state is still an equal superposition but with a 90° phase shift on the horizontal state now:

$$|P1\rangle = \frac{1}{\sqrt{2}}(i\,|x\rangle + |y\rangle)$$

A. If we are working in the $\{|x\rangle, |y\rangle\}$ basis, come up with the matrix that represents the action of the beam splitter on the two basis states. Find the eigenvalues and eigenvectors of the matrix.

B. Now, consider what would happen if we still had a symmetric beam splitter but instead of 50/50, we used a 60/40 beam splitter where 60% of the incoming photons are reflected and 40% are transmitted. Construct the matrix that represents the action of this beam splitter and show that it transforms the horizontal and vertical states, as expected.

Chapter 3

Introduction to Quantum Mechanics

Quantum mechanics is the physics describing nature at small scales (atom scale and smaller). Just as classical mechanics is governed by Newton's laws, quantum mechanics is governed by the Schrödinger equation. Since its birth, it has been crucial in explaining how the world works in ways classical physics cannot. From the structure of the atom, fusion in stars, superconductors, structure of DNA, and behavior of elementary particles, application of quantum mechanics has provided accurate predictions of experimental data. As a result, it is the accepted theory to explain physical processes. The interest now is to extend this framework to computing and see how it can be exploited to solve interesting and complex problems.

The concept of **wave–particle duality**, where particles can act like waves, and waves like particles, is at the forefront of quantum theory. For example, in the double-slit experiment from optics, you can think about a particle version where a beam of electrons are sent one by one through a double slit. If electrons are unobserved, we will see an interference pattern like for waves, indicating that particles like electrons have a wave-like nature. On the other hand, when a laser of a certain frequency is directed towards a piece of metal, electrons (which make up the metal) are emitted from the surface. This means that single particles of light called photons collided with each electron in the metal, thereby transferring their energy to the electron and allowing it to escape from the surface of the metal. In this sense, light which we often think of as a wave exhibits this particle-like property.

The mathematical background from Chapters 1 and 2 will now be combined to understand quantum mechanics which is a linear theory. This means that if A and B are the solutions to the Schrödinger equation, then $A + B$ is also a solution. Quantum mechanics relies on linear algebra, and the solution to Schrödinger's equation represents the probability distribution for a quantum object like a qubit. In this chapter, we will study the examples of light polarization and spins in magnetic fields to demonstrate how quantum objects work. These are examples of how qubits could be implemented and will aid our understanding how quantum information works.

Schrödinger's Cat

Erwin Schrödinger, a key physicist who helped develop fundamental results in quantum theory, came up with the Schrödinger equation, which provides the "wave mechanics" approach to quantum mechanics. The Schrödinger equation is a differential equation used to solve for the wavefunction (or the state) of a quantum system. This wavefunction is a complex vector which contains all the necessary spatial information about a quantum state and how it will change in time. Solving for the wavefunction is analogous to finding the superposition or sum of possible states that the quantum system could be in. Then, based on boundary and initial conditions from the physical system, the scaling factors can be calculated. Boundary conditions are spatial, while initial conditions are based on time.

To motivate the peculiarity of quantum mechanics, we will discuss an apparent paradox and misunderstanding of quantum superposition in the Copenhagen interpretation. In 1935, Schrödinger developed a thought experiment in discussion with Albert Einstein [43]:

> There is cat locked in a steel chamber along with a device that follows its movements. On the wall, there is a Geiger counter[1] affixed on the wall with a small amount of a radioactive substance. The radioactive substance has a 50% chance of decaying in the course of 1 hour or a 50% chance of not decaying. In the case that it decays, the tube of the Geiger counter will discharge and a relay will release a hammer that shatters a small flask of hydrocyanic acid which will poison the cat.

---

[1] A Geiger counter is a device that measures radioactive particles, such as alpha, beta and gamma.

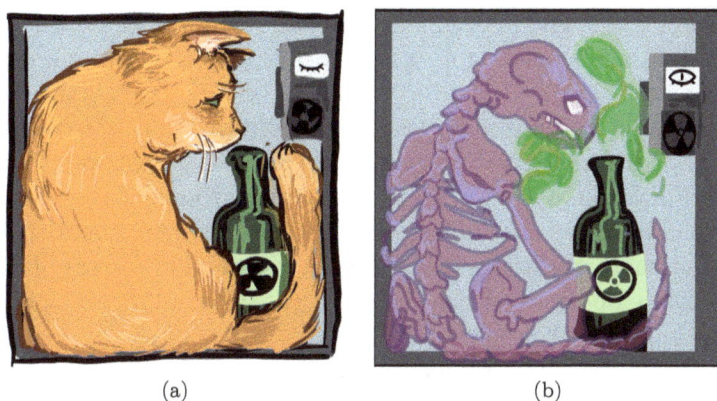

Figure 3.1. (a) Geiger counter is not triggered and the cat is alive; (b) Geiger counter is triggered and the cat is poisoned.

The thought experiment can be visualized in Fig. 3.1. We can consider $|0\rangle$ to represent the state that the radioactive substance does not decay and $|1\rangle$ to be represent the state that radioactive substance decays. Until the radioactive substance is measured, it exists in a superposition of the states $|0\rangle$ and $|1\rangle$. Since the radioactive substance is a quantum system, there is some probability to measure either $|0\rangle$ or $|1\rangle$. However, we will only obtain one answer from measuring the state. If we leave the state unchanged and keep measuring it, we will continue to get the same answer, as we should since we did not change anything. So, in this sense, there is no way for us to know the probability distribution from a single measurement. The only way we could obtain the distribution is to have multiple radioactive substances that are prepared in the same situation and repeatedly measure each system.

For now, let us assume that someone has already done the hard work of gathering statistics and figured out that this particular radioactive substance has a 50% chance of decaying within the course of the hour, and thus a 50% chance of not decaying. Let us also assume, as was believed by some physicists at the time, that only *conscious* observers can cause the superposition state to "collapse" to a definite event. Then, the whole system in the steel chamber is quantum, and the cat has a 50% chance of being alive and a 50% chance of being dead. Only when an observer opens the door will the fate of the cat be determined. So, the question that is posed by this famous thought experiment is: "What/who decides if a quantum system stops existing as a superposition and becomes a definite state?"

In fact, the resolution to the paradox is that not *only* conscious observers can cause the quantum superposition to collapse, but *any* interaction of the quantum state with its environment can cause the state to collapse to a definite event [14]. In this steel chamber, there is the radioactive substance, the Geiger counter and the cat. The fate of the cat is determined way before a conscious observer opens the door to look.

To summarize, the radioactive substance is a quantum system that exists in an equal superposition of decaying and not decaying but with the cat and the Geiger counter in the box, this indeterminate state has probabilistically collapsed. Thus, the substance has either decayed or not so that the Geiger counter *is* not triggered and the cat remains alive OR the Geiger counter is triggered and the cat is poisoned, each with 50% chance of occurring. In other words, it has become a definite or *classical* event. In general, the thought experiment is used to illustrate that an observation or measurement is not necessarily performed by a so-called "conscious" observer. It can be thermal effects in the environment or a cat that does not know anything about radioactivity!

Checkpoint Exercises

1. A quantum state can exist in a superposition which can be described by linear algebra. This is because quantum mechanics is a linear theory similar to classical Maxwell equation.

   a. Based on what we have learned so far about superposition, what are the two independent states (or independent vectors) for the radioactive substance?

   b. What are the scaling factors for those independent states? Write the superposition state for $|\psi\rangle$. (*Hint*: Start with what we know about the probabilities of getting $|0\rangle$ and $|1\rangle$ and then normalize the state.)

2. We have learned now that quantum states can be disturbed/measured by any interaction they have with the outside the world.

When does the superposition state of the radioactive particle collapse? (*Hint*: Think about everything that is inside the steel chamber with the radioactive particle.)

3.1 Measurement of Quantum States

In the Schrödinger's cat example, we only briefly discussed measurement or observation of quantum states and the idea of the superposition state probabilistically collapsing when it is observed. This observation or measurement can be a conscious observer like us, conducting an experiment, and interacting with the quantum object to learn about it. It could also be the environment itself! Quantum states are extremely fragile and retain their superposition in well-isolated and/or sub-Kelvin temperatures (close to absolute zero[2]). This means that we should be aware of thermal effects, which always exist at finite temperature. If the temperature is large enough (e.g. room temperature), the superposition could collapse (e.g. superconducting qubits). This is one of the reasons why we do not encounter quantum behavior in our macroscopic experience. The other reason is dissipation. It is important to mention that quantum behavior can be observed at millimeter scale when temperature is low enough (compared to qubit energy) and dissipation is removed.

Another reason why we do not see quantum effects in everyday objects is because quantum objects have energy quantization which exists at atomic and subatomic distances characterized by the unit angströms (Å) that is of the order of 10^{-10} m. Energy quantization means that quantum objects can only have certain energies, and in between two energies, there is nothing (only observable if temperature is smaller than quantization energy). This is something that we cannot see in a macroscopic object since they have so many of these atoms bonding together. As a result, the energy spectrum becomes extremely dense and appears pretty much continuous, as shown in Fig. 3.2.

---

[2]Absolute zero or 0 K (−459°F) is, as it sounds, the lowest absolute temperature that can be achieved, where essentially all motion of particles stops. While absolute zero can never be reached, experimenters have gotten very close. Temperatures around 10 mK can be routinely achieved using dilution refrigerators.

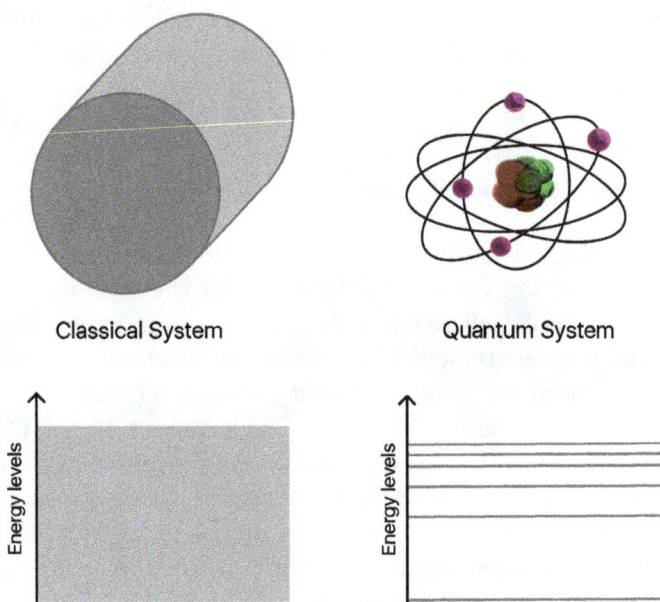

Figure 3.2.   Quantum objects like an atom have discrete energy levels, while classical objects like a metal rod appear to have continuous energy levels.

In this section, we discuss in detail *how* a quantum state is measured assuming that it is well isolated, and we are experimenters trying to study it. The key takeaway should be understanding what the **measurement basis** or **computational basis** is and how **projective measurement** work.

3.1.1   Spins in Magnetic Fields

One important physical example to demonstrate how the choice of measurement basis affects measurement of quantum systems is the Stern–Gerlach experiment that we briefly discussed in the introduction. We know that atoms and subatomic particles such as electrons, protons and neutrons have some intrinsic properties like charge and weight. The Stern–Gerlach experiment was critical in showing the existence of another intrinsic property called spin. It's difficult to explain what spin is exactly because there is no direct analogue to classical physics, but it exists nonetheless!

In the Bohr version of an atom, there is a dense nucleus in the middle and electrons orbiting around this nucleus in specific orbits corresponding to the energy levels of the atom. Since electrons are moving charged particles, a current is induced around this "orbit" and therefore a magnetic field. The Bohr version of the atom can also be considered as a "planetary model" of the atom where electrons move around in quantized or specific energy levels. Although this is an intuitive understanding from a classical perspective, this description was not correct because electrons aren't really in orbits but rather in a "cloud" around the nucleus. This means that there's no way to know where the electron is at any given point in time. Bohr's model is accurate for the hydrogen atom which only contains one electron, but for atoms with multiple electrons, it is merely an approximation!

Experimental Details

We can think of the electron in the orbit as a loop of current I and area A produces a magnetic moment denoted μ:

$$\mu = IA$$

The current due to a charged particle like an electron is simply $I = \frac{e}{2\pi r}v$, where e is the charge of the electron moving around a circular orbit of circumference $2\pi r$ with velocity v. Then, the magnetic moment can be expressed as

$$\mu = v\frac{e}{2\pi r}\pi r^2 = \frac{er}{2}v$$

Now, we will make one final substitution for the quantity known as orbital angular momentum, $L = mvr$, where m is the mass of electron in this case. This orbital angular momentum for an electron moving around an orbit is analogous to the earth revolving around the sun. So, we can finally express the magnetic moment as[3]

$$\vec{\mu} = \frac{e}{2m}\vec{L}$$

---

[3]Note that angular momentum is actually defined as $\vec{L} = \vec{r} \times m\vec{v}$. This is a vector quantity that contains x-, y-, z-components and is computed by taking the cross-product of the radius vector, $\vec{r}$ and velocity vector, $\vec{v}$. To simplify the above derivation, we assumed that the velocity and the radius vector of the loop were perpendicular so that the magnitude of $\vec{L}$ was simply $|\vec{L}| = mvr\sin 90° = mvr$ [11].

Figure 3.3.   Visualization of the spin of a quantum particle.

However, the earth also rotates around its own axis, so keeping with the analogy, the electron could also rotate around its own axis and possess a property known as intrinsic angular momentum or **spin**, as is heuristically represented in Fig. 3.3.

Spin can be expressed as[4]

$$\vec{\mu}_s = g \frac{e}{2m} \vec{S}$$

We are skipping the detailed derivation of this equation, but that information is encapsulated in the additional dimensionless factor g known as the g-factor. This is as far as the analogy of an electron being a sphere that rotates around its own axis can go. This is because fundamental particles like electrons that cannot be subdivided into smaller components are considered to be *point-like*, yet they still possess this property of spin. We may, for our own intuition, think of these quantum particles as tiny spheres rotating about their own axis. However, this is only a heuristic description and we know this is not correct.

Particles that possesses a spin magnetic moment (like electrons) experience a torque when an external magnetic field is applied, which leads to a rotation about the origin. The vectors would trace out equal cones opening in different directions as presented in Fig. 3.4.

---

[4]Again, spin is intrinsic angular momentum and is a vector quantity that contains x-, y-, z-components.

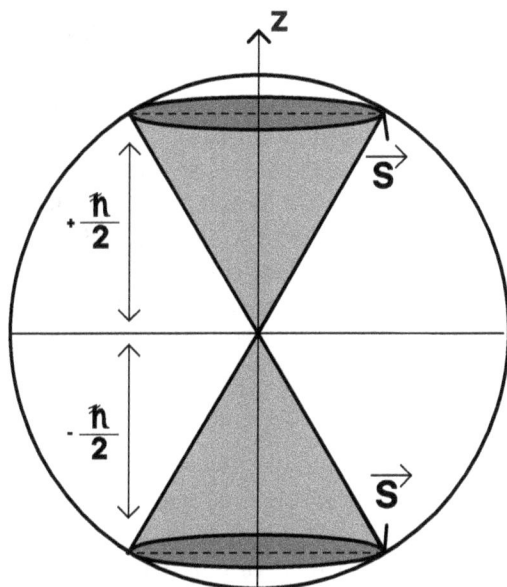

Figure 3.4. Spin $\frac{1}{2}$ particles in an external magnetic field.

The electron can be considered as a spinning top or gyroscope that creates a magnetic moment, $\vec{S}$. There are two cones traced out because the spin of an electron could be either "spin-up" or "spin-down." If we only took the z-component of $\vec{S}$, S_z, then we could measure the spin to be either $+\frac{1}{2}\hbar$ or $-\frac{1}{2}\hbar$, as shown in the figure.[5] If we tried to measure x- or y-components of $\vec{S}$, we would also measure $+\frac{1}{2}\hbar$ or $-\frac{1}{2}\hbar$. Other particles like protons and neutrons are also spin $\frac{1}{2}$ particles, so they also have two orientations along any axis: "spin-up" or "spin-down."

The original setup of the Stern–Gerlach experiment measured the z-component of spin for neutral silver atoms going through a non-uniform magnetic field which is now known as a Stern–Gerlach apparatus or SG apparatus [25]. Neutral atoms are used because it can be very difficult to observe the quantized spin effects when working

---

[5]The constant $\hbar$ is known as reduced Planck's constant and is fundamental to quantum mechanics and has units of angular momentum (kg · m^2/s).

with charged particles like electrons alone.[6] So, how could these neutral silver atoms then be used to show the existence of spin? Well, a silver atom is composed of 47 electrons, 47 protons and somewhere between 60 and 62 neutrons (depending on the isotope used). Protons and neutrons are much heavier than electrons,[7] so they have negligible contribution to the magnetic moment that we derived, which was inversely proportional to mass. Now, with a quick look at the electron configuration of silver, we find that the outermost shell contains only one electron which gives the total contribution to the magnetic moment of the atom.

Measurement of Spins

Let's suppose we have a source of silver atoms going through one by one into the SG apparatus so that we can measure their spin [26]. Its outer electron comes in at some arbitrary spin orientation, $|\psi\rangle$, and goes through magnetic field oriented mostly in the z-direction which orients the spin axis of the electron along what we consider the z-direction and thus we can observe the possible outputs $|\uparrow\rangle$ or $|\downarrow\rangle$, as shown in Fig. 3.5.

These states $|\uparrow\rangle$ or $|\downarrow\rangle$ are the states that make up the z computational basis. The electron is initially in an arbitrary superposition state:

$$|\psi\rangle = \alpha\,|\uparrow\rangle + \beta\,|\downarrow\rangle$$

Once the electron goes through the SG apparatus, it will either collapse to the $|\uparrow\rangle$ or $|\downarrow\rangle$ state with a 50% chance of either state to occur. If we then were to send the resulting state through an SG apparatus oriented in the x-direction, then we would measure $|\rightarrow\rangle$ or $|\leftarrow\rangle$ with a 50% chance of either state occurring. The total probability of measuring either of these states is only 25% of the original. It is clear that the choice of basis greatly influences the results of measurements. Basically, we can think about the basis states as the

---

[6]An electron in the presence of a magnetic field experiences a Lorentz force given as $q\vec{v} \times \vec{B}$, where q is the charge of the electron. As a result, their trajectory will be curved and they tend to have large deflections when traveling through a magnetic field.

[7]The mass of protons and neutrons is about 2000 times more than the mass of an electron!

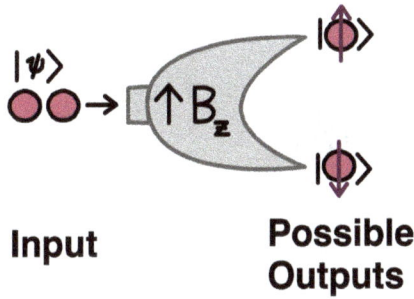

Figure 3.5. The input to the SG apparatus is a free electron, $|\psi\rangle$, with arbitrary spin orientation and the outputs are the computational basis states for the z-components, $|\uparrow\rangle$ or $|\downarrow\rangle$.

axes which provide us with orientation in the space. Physically, it would be the alignment of our measurement apparatus. In the case of the SG apparatus, it is the magnet that is set up to measure some specific component of the spin.

Now, because of the quantization of quantum states, when a measurement is performed, the result can only be one of the basis states. This is because the measurement apparatus can only detect those basis states, that is how it is set up. However, prior to measurement, the quantum state is in superposition, and the result when measured is *probabilistic*. The probability of getting $|\uparrow\rangle$ is $|\alpha|^2$ and $|\downarrow\rangle$ is $|\beta|^2$, and these two values must add up to one:

$$|\alpha|^2 + |\beta|^2 = 1$$

This is the physical meaning of the normalization condition we introduced in Chapter 2. Furthermore, the states $|\uparrow\rangle$ and $|\downarrow\rangle$ are orthogonal, so $\langle\uparrow|\downarrow\rangle = 0$, so they do not overlap with each other. The conclusion here is that the basis must be chosen carefully because once a quantum state is measured, the state prior to measurement is destroyed!

Checkpoint Exercises

1. If a source of silver atoms is sent through the SG apparatus where its magnetic field is oriented in y-direction, what would we see on the screen?
2. Random electron spins are sent through two consecutive SG apparati both with magnetic fields oriented in the z-direction, what is

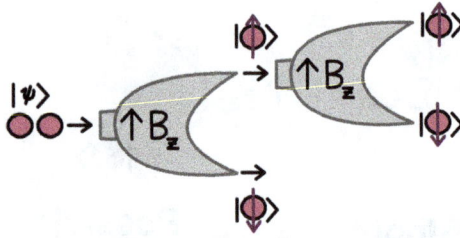

Figure 3.6.   Figure for Question 2.

the output of the second apparatus and what is the probability of getting that output (Fig. 3.6)?

3. Random electron spins are sent through the following SG apparati. What is the final output? What is the final probability of getting either "spin-up" or "spin-down," considering the entire sequence shown in Fig. 3.7?

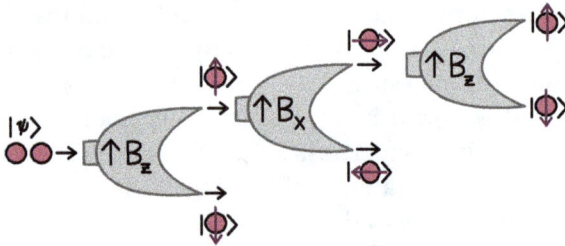

Figure 3.7.   Sequence for Question 3.

3.1.2   Light Polarization

What we see as visible light is just electromagnetic waves that are oscillating in spacetime. As we know, in vacuum, these waves move at the speed of light, $c = 2.998 \times 10^8$ m/s, but in other media like water or glass, they move at a slower speed. However, there is no media where the waves move faster than c. Light also has other properties. For example, polarization is a property of light that defines the geometrical orientation of the oscillation which is perpendicular to the direction of propagation or the direction the wave is moving. If we consider a single-mode laser, which emits coherent light that can be focused to a tightly contained spot in space, the laser will

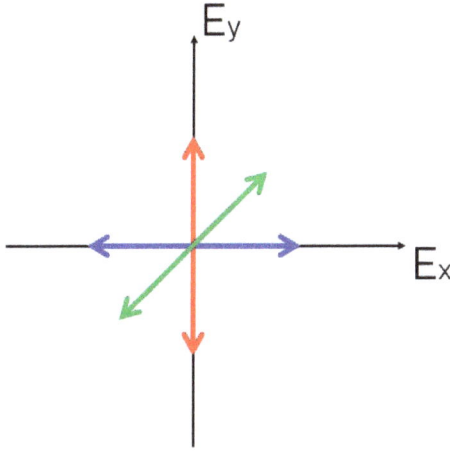

Figure 3.8. Red arrow denotes vertically polarized light. Blue arrow denotes horizontally polarized light. Green arrow indicates linearly polarized light at a 45° angle.

be shining light outward, say in the z-direction, while the electric field (which may be described by some sinusoidal wave) is oscillating in the xy-plane. A field that oscillates in the horizontal direction (x-axis) is known as *horizontally polarized light,* and we will denote the unit vector of this direction as $|\rightarrow\rangle = \binom{1}{0}$ or $|H\rangle$ for horizontal. A field that oscillates in the vertical direction (y-axis) is known as *vertically polarized light* and we will denote the unit vector of this direction as $|\uparrow\rangle = \binom{0}{1}$ or $|V\rangle$ for vertical.

Light can also be linearly polarized at an angle. For example, we can have linearly polarized light at 45°, denoted $|\nearrow\rangle$ ($|D\rangle$ meaning diagonal) which can be understood as an *equal superposition* of horizontally and vertically polarized light. The orthogonal vector to $|\nearrow\rangle$ is $|\searrow\rangle$ ($|A\rangle$ for anti-diagonal):

$$|\nearrow\rangle = |D\rangle = \frac{1}{\sqrt{2}}(|\rightarrow\rangle + |\uparrow\rangle) = \frac{1}{\sqrt{2}}\binom{1}{1}$$

$$|\searrow\rangle = |A\rangle = \frac{1}{\sqrt{2}}(|\rightarrow\rangle - |\uparrow\rangle) = \frac{1}{\sqrt{2}}\binom{1}{-1}$$

The reason why we have $\frac{1}{\sqrt{2}}$ scaling the two horizontal and vertical unit vectors is because we also want $|\nearrow\rangle$ to be a unit vector in this

45° direction, and we know that unit vectors must have magnitude of 1:

$$\left|\left|\nearrow\right\rangle\right| = \sqrt{\langle\nearrow|\nearrow\rangle} = \sqrt{\left(\frac{1}{\sqrt{2}}\right)^2 + \left(\frac{1}{\sqrt{2}}\right)^2} = \sqrt{\frac{1}{2} + \frac{1}{2}} = 1$$

We can also generate circularly polarized light, in which the electric field oscillates in circular fashion clockwise, $|\circlearrowright\rangle$), or counterclockwise, $|\circlearrowleft\rangle$). Circularly polarized light in the clockwise direction may also be denoted as $|R\rangle$, implying "right-handed," while the light in the counterclockwise direction is denoted as $|L\rangle$, implying "left-handed" as represented in Fig 3.9.

Circular polarizations can occur when we shift the phase of the electric field by ±90°. So, we would have the horizontal and vertical components in equal superposition but with a phase difference of 90°. This phase difference occurs when we multiply $|\uparrow\rangle$ state by the imaginary number i, which we learned in Chapter 1 represents a 90° rotation:

$$|\circlearrowright\rangle = |R\rangle = \frac{1}{\sqrt{2}}(|\rightarrow\rangle - i|\uparrow\rangle) = \frac{1}{\sqrt{2}}\begin{pmatrix} 1 \\ -i \end{pmatrix}$$

$$|\circlearrowleft\rangle = |L\rangle = \frac{1}{\sqrt{2}}(|\rightarrow\rangle + i|\uparrow\rangle) = \frac{1}{\sqrt{2}}\begin{pmatrix} 1 \\ i \end{pmatrix}$$

Light polarization is widely used to measure materials and molecules because they tend to rotate the polarization of light that

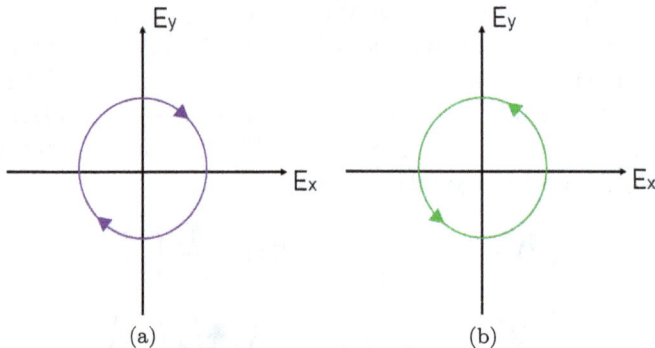

Figure 3.9. (a) Right-handed polarization is in the clockwise direction; (b) left-handed polarization is in the counterclockwise direction.

passes through them. Substances such as sugar water, collagen and insulin are considered *optically active* because they change the polarization of light that passes through them. So, light polarization has nothing "inherently" quantum about it, but it is light itself which has a quantum description. Classically, we know that light is electromagnetic waves, but its quantum description is the photon which is an indivisible bundle of energy. We may think of the photon as an infinitesimal blip of light that does not extend over any distance but has some polarization. Single photon emitters have been researched and can be used along with optical components such as polarizers, mirrors, beam splitters, and crystals to manipulate the polarization of the photon and thus encode information in it. This forms the basis for optical quantum computing.

Checkpoint Exercises

1. Express the vertical polarization state in the $\{|\nearrow\rangle, |\searrow\rangle\}$ basis.
2. Express the horizontal polarization state in the $\{|\circlearrowright\rangle, |\circlearrowleft\rangle\}$ basis.

Measurement with Light

The first step to understanding measurement is to have an intuitive idea about the computational basis. Suppose that a light wave is coming in at a 45° angle, which we can denote as $|\nearrow\rangle$, but we only have polarizing filters positioned $|\uparrow\rangle$ and $|\rightarrow\rangle$. Well, we know that $|\nearrow\rangle$ is equally composed of $|\uparrow\rangle$ and $|\rightarrow\rangle$. Applying either polarizer will allow *that* component through. Note in Fig. 3.10 that after applying a polarizer and analyzer in case 1, the information about the original wave, $|\nearrow\rangle$, is gone, and we only have the $|\uparrow\rangle$ component. Of course, if we run this wave through the $|\rightarrow\rangle$ polarizer, we would expect no light to go through because there is no $|\rightarrow\rangle$ component in the $|\uparrow\rangle$ light wave.

However, we can change that second polarizer to $|\nearrow\rangle$ instead of the $|\rightarrow\rangle$ polarizer for the third case in the figure. In this case, we know that $|\uparrow\rangle$ is equally composed of $|\nwarrow\rangle$ and $|\nearrow\rangle$. Thus, we would observe the $|\nearrow\rangle$ component. Once again, information about the incoming $|\uparrow\rangle$ wave is destroyed with this choice of measurement bases.

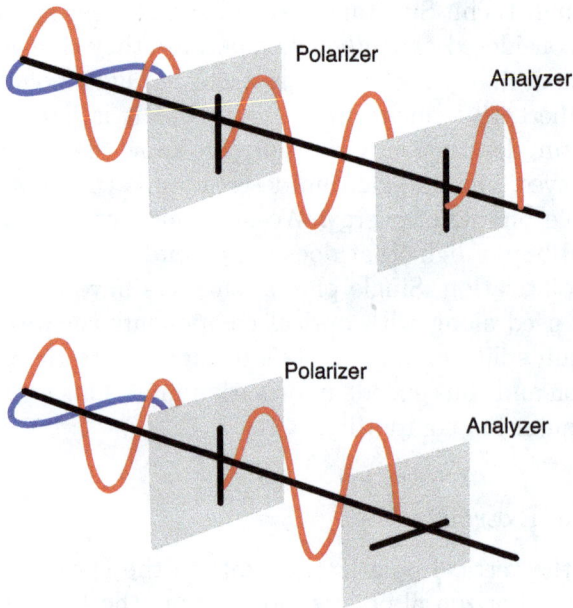

Figure 3.10. Light waves going through polarizers.

Recall that waves have some amplitude and frequency associated with them. When the $|\nearrow\rangle$ goes through the $|\uparrow\rangle$ polarizer, the amplitude of the wave is not the same. The resulting wave after the polarizer is 50% of the original amplitude of the $|\nearrow\rangle$. If we were to send the $|\uparrow\rangle$ wave through a $|\nearrow\rangle$ polarizer than once again, we would get 50% of the original amplitude of the $|\uparrow\rangle$, which is 25% of the first $|\nearrow\rangle$ wave.

So, how does this relate to qubits? Light obeys Maxwell's rules of electromagnetism which is a linear theory. Qubits obey quantum rules through the Schrödinger equation which is also linear. This is in contrast to Newton laws of motion for particles which is a non-linear theory. Hence, there is a close connection between light waves and qubits! The takeaways are that the choice of polarizing filters sets the **computational basis** which are the results of measuring or observing the system. The filter causes the original state to *collapse* to one of the basis states. It is always best to choose two directions that do not overlap with each other (orthogonal), like the $|\uparrow\rangle$ and $|\rightarrow\rangle$ polarizers, which you can always distinguish unambiguously from each other. Now, we see that the act of measurement is destructive because it forces the original state into one of the basis states. However, this is a necessary evil because otherwise we would not get any information!

Checkpoint Exercise

Consider light polarized in the 45° direction, passing through a $|\uparrow\rangle$ polarizer, then another $|\nearrow\rangle$ polarizer and finally a $|\rightarrow\rangle$ polarizer. Label the quantum state of the photon before passing through each polarizer and the effect of the polarizers on the amplitude of the photon state.

3.2   Postulates of Quantum Mechanics

In mathematical terms, a postulate is a statement that is taken to be true. In the case of quantum mechanics, these postulates were summarized after many aspects of the behavior of quantum states were tested theoretically and experimentally. The goal is to summarize the fundamental elements that make up the theory [26]:

1. The state of a quantum system is completely specified by a function, ψ, the wavefunction, that can be represented as a complex vector $|\psi\rangle = \begin{pmatrix} \alpha \\ \beta \end{pmatrix}$. The complex probability amplitudes are α and β. The squared magnitudes of these quantities, $|\alpha|^2$ and $|\beta|^2$, represent the probability of finding the system at a particular location after some time. For example, in the Schrodinger cat example, the radioactive particle has a 50% chance of decaying or not decaying. The corresponding probability amplitude would be $\frac{1}{\sqrt{2}}$ which when squared yields $\frac{1}{2}$. As a result of this probabilistic interpretation, $|\alpha|^2 + |\beta|^2 = 1$, or in other words, $|\psi\rangle$ must be a unit vector.
2. Every observable in classical mechanics, such as position, velocity, momentum, and energy, corresponds to a linear operator in quantum mechanics.

3. The measurement results of any observable are the eigenvalues of the operator that represents it. For example, in the Stern–Gerlach experiment, spin is the observable or quantity that is being measured. Eigenvalue $+\hbar/2$ corresponds to measuring the state $|\uparrow\rangle$ and eigenvalue $-\hbar/2$ corresponds to measuring the state $|\downarrow\rangle$. Any arbitrary spin state that goes through the apparatus will always yield either $|\uparrow\rangle$ or $|\downarrow\rangle$.

Homework 3

Let's consider an optical setup called a Mach–Zehnder Interferometer, which is set up on Quantum Flytrap, as shown in Fig. 3.11. You may click on experimental setups → Mach–Zehnder Interferometer to use the pre-made setup. For more information on Quantum Flytrap, refer to Appendix B.

This setup is used to determine a phase shift between a photon whose path is split in two using a beam splitter. In one path (transmitted in this case), the photon encounters some medium or sample, like glass, which may alter its phase. The photon is then

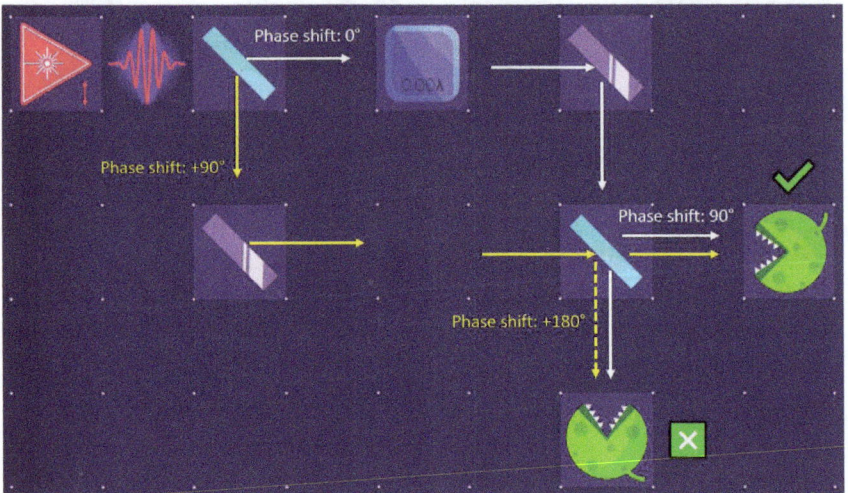

Figure 3.11. Path and corresponding phase shifts for the initially split photon. The red triangle is a laser source, the gray blocks are mirrors, the blue blocks are beam splitters and the curved square plate between the beam splitter and the mirror on the top row is a piece of glass whose refractive index can be changed.

reflected off of a mirror and travels to second beam splitter before being detected. In the other path (reflected in this case), the photon reflects off a mirror and travels to the second beam splitter before being detected.

1. Without altering the phase of the glass yet, send a couple of photons in using the "Loop" feature. What do you observe?
2. Now, change the phase of the glass to 0.5. What do you observe now?
3. Recall that beam splitters can introduce a phase shift in reflected light. In this virtual optical table, the beam splitter is "symmetric," implying that on each side light is reflected, a 90° phase shift is induced, and both of these phase shifts add up to 180° or π, as we discussed before for a non-symmetric beam splitter. Use the backward and forward buttons to observe the phase changes in the split photon as it goes through its path and verify the arrow directions and phases indicated in Fig. 3.11.
4. Now, perform a similar analysis on the case when the phase of the glass is 0.5.
5. What occurs to the photon path when the glass is set to a 0.25 phase shift?

Chapter 4

Single Qubit Representation and Measurement

In this chapter, we develop a visual representation of a single qubit based on what we have learned so far. A single qubit is a two-level quantum system that can be represented as a complex 2D vector. We will show that the 2D complex vector, can be mapped to a 3D vector so that we can visualize it.

So far, we have represented the qubit as a superposition of the unit vectors $|0\rangle$ and $|1\rangle$:

$$|\psi\rangle = c_1 |0\rangle + c_2 |1\rangle$$

where the scaling coefficients, c_1 and c_2, may be complex. These complex coefficients must satisfy the following normalization condition for $|\psi\rangle$ to be a unit vector:

$$|c_1|^2 + |c_2|^2 = 1$$

If we rewrite the complex coefficients in terms of Euler's formula as we learned in Chapter 1, then we recover an intuitive connection between representing $|\psi\rangle$ as a 2D complex vector and as a vector on a unit sphere called the Bloch sphere, whose location can be defined in terms of two angles using spherical coordinates. This representation will allow us to visualize how the state of the qubit changes when we operate on it with quantum gates.

The quantum gates we will cover include the Pauli matrices, rotation matrices and more. We will also go into more detail about how measurement works and the importance of choosing the right computational basis to measure quantum states, generating probability distributions, and calculating expected values.

4.1   Bloch Sphere

To motivate the discussion of mapping the complex vector $|\psi\rangle$ to a point on a unit sphere, we use the analogy of thinking of the qubit as a globe where north pole is $|0\rangle$ and the south pole $|1\rangle$, as shown in Fig. 4.1.

The arbitrary state of the qubit may be anywhere on the surface of this globe and the scaling factors c_1 and c_2 can be translated to two angles. This is analogous to specifying longitude and latitude to give one's position on Earth.

To mathematically see this mapping, we refer back to our superposition state which we used to represent an arbitrary state for the qubit:

$$|\psi\rangle = c_1 |0\rangle + c_2 |1\rangle$$

Figure 4.1.   Analogy of the globe with a single qubit.

We know that coefficients c_1 and c_2 are complex numbers and can be expressed in terms of Euler's formula,

$$c_1 = r_1 e^{i\varphi_1}$$

$$c_2 = r_2 e^{i\varphi_2}$$

Here, r_1 and r_2 represent the radii of the circles formed by the complex numbers, and φ_1 and φ_2 represent the angles from the real axis for each of the complex numbers. Let's plug these into the superposition for $|\psi\rangle$:

$$|\psi\rangle = r_1 e^{i\varphi_1} |0\rangle + r_2 e^{i\varphi_2} |1\rangle$$

Note here that we still have four unknowns, $r_1, r_2, \varphi_1, \varphi_2$, just as we would have had if we tried to solve for the complex coefficients in standard form. Now, let's simplify the equation by dividing both sides by $e^{i\varphi_1}$:

$$|\psi\rangle e^{-i\varphi_1} = r_1 |0\rangle + r_2 e^{i\varphi_2 - \varphi_1} |1\rangle$$

Let's define the relative angle of rotation in the complex plane to be $\varphi = \varphi_2 - \varphi_1$ and cut down on one unknown angle:

$$|\psi\rangle e^{-i\varphi_1} = r_1 |0\rangle + r_2 e^{i\varphi} |1\rangle$$

Now, we see that the quantum state vector must have a magnitude of 1, so let's see if $|\psi\rangle e^{-i\varphi_1}$ also has a magnitude of 1:

$$|\langle\psi|\psi\rangle e^{-i\varphi_1}|^2 = |\langle\psi|\psi\rangle|^2 (e^{-i\varphi_1} e^{i\varphi_1})^2 = ||\psi\rangle|^2 = 1$$

The $e^{-i\varphi_1}$ factor is known as a *global phase factor*, and when we take its magnitude we need to multiply by its complex conjugate and which gives a magnitude of 1. This means that global phase factors do not change the magnitude of the quantum state. Now, we still have the familiar normalization condition:

$$||\psi\rangle|^2 = |r_1|^2 + |r_2 e^{i\varphi}|^2 = 1$$

Here, r_1 and r_2 are real numbers. As we just saw, when we take the magnitude of a global phase factor like $e^{i\varphi}$, we will get 1. So, we can reduce this to the following:

$$r_1^2 + r_2^2 = 1$$

This is the familiar Pythagorean identity, so we can define $r_1 = \cos\theta$ and $r_2 = \sin\theta$. We have now one less unknown because of the normalization condition. So, we only need to specify the relative angle φ and the angle θ in order to specify our quantum state $|\psi\rangle$. Let's do a quick check to identify our angle ranges. If we want $|\psi\rangle = |0\rangle$, then we want $\theta = 0$, but if we want $|\psi\rangle = e^{i\varphi}|1\rangle$, then $\theta = \frac{\pi}{2}$. So, we can represent $|\psi\rangle$ as

$$|\psi\rangle = \cos\left(\frac{\theta}{2}\right)|0\rangle + \sin\left(\frac{\theta}{2}\right)e^{i\varphi}|1\rangle$$

where $0 \leq \theta \leq \pi$ and $0 \leq \varphi \leq 2\pi$. We have reduced the amount of unknowns from 4 to 2 because of the definition of the *relative angle* $\varphi = \varphi_1 - \varphi_2$ and the normalization condition! Now, we have mapped the representation of a qubit to a unit sphere, as can be seen in Fig. 4.2, where we only need to specify θ and φ to visualize the location of the qubit.

The main takeaway from this section is that the vectors $|0\rangle$ and $|1\rangle$ form an **orthonormal basis** which means that $|0\rangle$ and $|1\rangle$ are orthogonal to each other and each is normalized. This basis is defined as the minimum number of vectors needed to describe the space that we are working with. So, if we have one qubit, we need two basis states to describe an arbitrary state $|\psi\rangle$. If we have two qubits, then we need four basis states to describe an arbitrary state so we have the

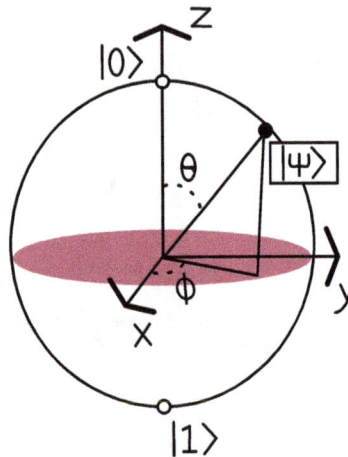

Figure 4.2.   Bloch sphere representation of a single qubit.

set $\{|00\rangle, |01\rangle, |10\rangle, |11\rangle\}$. For three qubits, we need 8 basis states $\{|000\rangle, |001\rangle, |010\rangle, |011\rangle, |100\rangle, |101\rangle, |110\rangle, |111\rangle\}$. For n qubits, we need 2^n basis states.

An interesting point is that for $n = 325$, we would have $2^{325} \approx 7 \times 10^{97}$ states. This is enough qubits to represent all the particles in the universe. Such state storage is not possible with a classical computer, which is why quantum computing can be so interesting for simulating complex systems. Any simulations of molecules done on classical computers ultimately relies on approximations in order to reduce the complexity of the problem. In turn, we only get approximate solutions to such problems because solving the full problem has exponential complexity. However, there is a catch: we only have access to 325 outputs or measurements. We cannot access the massive amount of stored information. This is where smart algorithms come in to evolve such large states such that when you collapse this large state by measuring it, you can get the answer you are looking for.

4.2 Controlling Qubits

We will discuss some important single qubit quantum gates in this section, implement them using Qiskit quantum circuits to visualize what happens on the Bloch sphere, and calculate the eigenvalues and eigenvectors for these gates. See Appendix A for more information to set up Qiskit. Keep in mind that quantum gates are square matrices which may contain complex elements. We represent our quantum states with vectors and these matrices rotate our qubit but preserve the dimensionality of the state.

4.2.1 Pauli Matrices

A set of important operations we will apply to our qubits are known as the Pauli matrices or Pauli gates. They are 2×2 Hermitian matrices that represent rotations around the Bloch sphere and form the basis for mathematically representing spin. They are named after the Austrian theoretical physicist Wolfgang Pauli, who provided the basis for the theory of spin.

These matrices/gates are listed as follows:

$$\mathbf{X} = \begin{pmatrix} 0 & 1 \\ 1 & 0 \end{pmatrix} \quad \mathbf{Y} = \begin{pmatrix} 0 & -i \\ i & 0 \end{pmatrix} \quad \mathbf{Z} = \begin{pmatrix} 1 & 0 \\ 0 & -1 \end{pmatrix} \quad \mathbf{I} = \begin{pmatrix} 1 & 0 \\ 0 & 1 \end{pmatrix}$$

As we have seen by now, it is very important to know the eigenvalues and eigenvectors of the operators that we work with. Using the techniques we have learned, we can easily see that all of the Pauli spin matrices except for the identity matrix have the same eigenvalues: $+1$ and -1. The eigenvalues are real, so all of these matrices correspond to an observable quantity, which is the spin components in this case. Knowing the eigenvalues, we can show that the eigenvalues are as follows:

$$X : |u_1\rangle = \frac{1}{\sqrt{2}} \begin{pmatrix} 1 \\ 1 \end{pmatrix} \quad |u_2\rangle = \frac{1}{\sqrt{2}} \begin{pmatrix} 1 \\ -1 \end{pmatrix}$$

$$Y : |v_1\rangle = \frac{1}{\sqrt{2}} \begin{pmatrix} 1 \\ i \end{pmatrix} \quad |v_2\rangle = \frac{1}{\sqrt{2}} \begin{pmatrix} 1 \\ -i \end{pmatrix}$$

$$Z : |w_1\rangle = \begin{pmatrix} 1 \\ 0 \end{pmatrix} \quad |w_2\rangle = \begin{pmatrix} 0 \\ 1 \end{pmatrix}$$

Qiskit Implementation: X-gate

We worked with the **X**-gate in the previous chapter and saw that we could flip the $|0\rangle$ state to $|1\rangle$ and vice versa using this matrix. Let's construct a one qubit quantum circuit in Qiskit and visualize how the $|0\rangle$ state is rotated on the Bloch sphere. In Qiskit, the default qubit state is $|0\rangle$, so we can generate a circuit that contains one quantum register to store the qubit we are working with and one classical register to store the result of measuring that qubit if we were to measure it. As we can see, the Pauli **X**-gate performs the expected operation of flipping the qubit from $|0\rangle$ to $|1\rangle$, which on the Bloch Sphere is shown as a 180° rotation. You can check that if you apply the gate twice, you will end up back at $|0\rangle$.

```
from qiskit import *
from qiskit.quantum_info import Statevector
from qiskit.visualization import plot_bloch_multivector, plot_histogram

q = QuantumRegister(1) #Initialize Quantum Register
c = ClassicalRegister(1) #Initialize Classical Register for measurement results
circ = QuantumCircuit(q,c) #Create Quantum Circuit
circ.x(0) #Pauli X-gate applied
circ.draw('mpl')
```

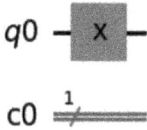

```
state = Statevector(circ)
print(state)
plot_bloch_multivector(state)
```

```
Statevector([0.+0.j, 1.+0.j],
            dims=(2,))
```

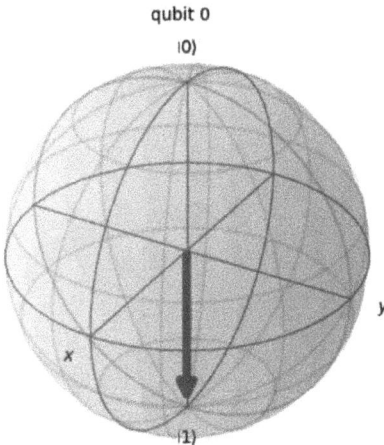

Qiskit Implementation: Y-gate

Now, let's try to implement the **Y**-gate in Qiskit and see what happens. The **Y**-gate looks similar to the **X**-gate in that they both have zeros across the diagonal elements. This means that we can expect that the qubit will be flipped from $|0\rangle$ to $|1\rangle$ and vice versa, but because there is an imaginary term, this means that the qubit will also pick up a phase of i or 90° when the gate is applied:

$$\mathbf{Y}|0\rangle = i|1\rangle$$

$$\mathbf{Y}|1\rangle = -i|0\rangle$$

Indeed, we observe that qubit was flipped, but we can see in the printed state vector that it picked up a phase. However, with the Bloch sphere, we will not be able to observe such global phases in the measurement (assuming projection on z-axis). For single qubits, only relative phases (when we have superposition of states) will be observable.

```
q = QuantumRegister(1) #Initialize Quantum Register
c = ClassicalRegister(1) #Initialize Classical Register for measurement results
circ = QuantumCircuit(q,c) #Create Quantum Circuit
circ.y(0) #Pauli Y-gate applied
circ.draw('mpl')
```

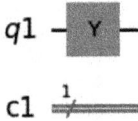

```
state = Statevector(circ)
print(state)
plot_bloch_multivector(state)
```

```
Statevector([0.+0.j, 0.+1.j],
            dims=(2,))
```

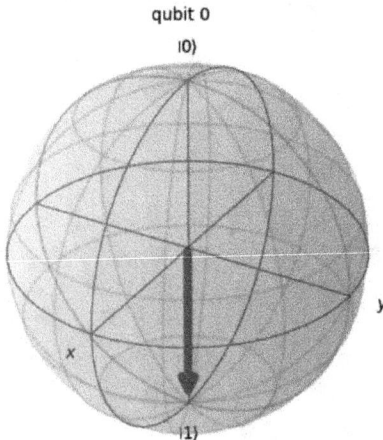

Qiskit Implementation: Z-gate

Finally, let us see the effect of using the **Z**-gate. In this case, we can tell from the matrix that **Z** will do nothing to $|0\rangle$ and contribute a global phase of $180°$ to $|1\rangle$ since there is a -1 in the second diagonal term. In terms of the Bloch sphere, we would not observe anything in Qiskit, but we can see from the Statevector simulator that when we start with the $|1\rangle$ state, we pick up a negative sign because the associated eigenvalue for the $|1\rangle$ state is -1.

```
q = QuantumRegister(1) #Initialize Quantum Register
c = ClassicalRegister(1) #Initialize Classical Register for measurement results
circ = QuantumCircuit(q,c) #Create Quantum Circuit
circ.x(0) #Pauli X-gate applied
circ.z(0) #Pauli Z-gate applied
circ.draw('mpl')
```

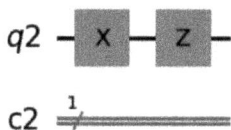

```
state = Statevector(circ)
print(state)
plot_bloch_multivector(state)
```

```
Statevector([ 0.+0.j, -1.+0.j],
            dims=(2,))
```

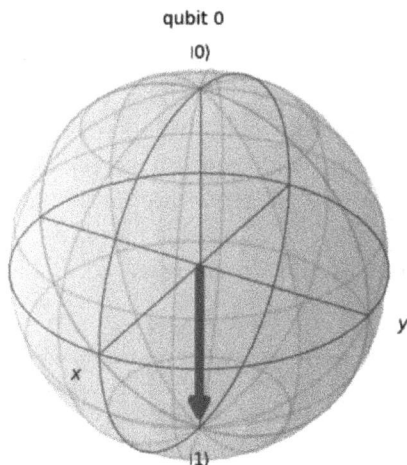

4.2.2   Rotation Gates

With the Pauli matrices, we observed 180° rotations and some global phase shifts which were not observable. Using arbitrary rotations which can be represented as unitary gates, we may fully rotate our qubit and move it to any point on the Bloch sphere.

Some intuitive rotations we may think about are how to rotate the qubit about the x-axis,[1] denoted $R_x(\theta)$, y-axis, denoted $R_y(\theta)$, or z-axis, denoted $R_z(\varphi)$, where θ is the angle the qubit is being rotated by from the z-axis and φ is the angle of rotation on the xy-plane, as shown in Fig. 4.2:

$$\mathbf{R}_x(\theta) = \begin{pmatrix} \cos(\theta/2) & -i\sin(\theta/2) \\ -i\sin(\theta/2) & \cos(\theta/2) \end{pmatrix} \quad \mathbf{R}_y(\theta) = \begin{pmatrix} \cos(\theta/2) & -\sin(\theta/2) \\ \sin(\theta/2) & \cos(\theta/2) \end{pmatrix}$$

$$\mathbf{R}_z(\varphi) = \begin{pmatrix} e^{-i\varphi/2} & 0 \\ 0 & e^{i\varphi/2} \end{pmatrix}$$

From the Bloch sphere, we know that $0 \le \theta \le \pi$ and $0 \le \varphi \le 2\pi$.

The easiest one to start with is $\mathbf{R}_z(\varphi)$. If the angle of rotation is $\varphi = 180°$, then we end up with the **Z**-gate. If we have $|0\rangle$ or $|1\rangle$ and we apply the $\mathbf{R}_z$ gate, then all that happens is that the state picks up a global phase factor which we said before is not something we can observe on the Bloch sphere. It does not affect the magnitude of the vector, and therefore, it won't affect the measurement result. So, if we measure $\mathbf{R}_z(\varphi)|0\rangle = e^{-\varphi/2}|0\rangle$ in the $\{|0\rangle, |1\rangle\}$ basis, we would just get $|0\rangle$. However, if the state was not along the z-axis but instead on xy-plane, then applying $\mathbf{R}_z(\varphi)$ would rotate the state around the xy-plane by angle φ.

Next, a rotation using $\mathbf{R}_y(\theta)$ would allow us to rotate our qubit in the xz-plane. If we take $|0\rangle$ state and rotate it by $\theta = \frac{\pi}{2}$, then we go from the z-axis to the x-axis fully and generate an equal superposition state in the $\{|0\rangle, |1\rangle\}$ basis. The same thing would occur with $\mathbf{R}_x(\theta)$ except now the rotations are on the yz-plane and that is why

---

[1]When we are rotating about some axis, say x-axis, it means imagine that with your right hand, you point your thumb along the x-axis and curl your fingers in the direction of the angle you are rotating, Positive rotations would be counter-clockwise, so your thumb would point outwards, and negative rotations would be clockwise, so your thumb would point inwards toward the page.

we have imaginary components in the off-diagonal elements similar to what we saw for Pauli **Y** matrix.

Qiskit Implementation of Rotation Gates

1. $\mathbf{R}_x(\theta)$ where $\theta = 30° = \frac{\pi}{6}$ rad:

As we can see on the Bloch sphere, $\mathbf{R}_x\left(\frac{\pi}{6}\right)$ rotated the qubit from the default state of $|0\rangle$ state on the yz-plane and the state vector is given in the printed array.

```
from math import pi
```

```
q = QuantumRegister(1) #Initialize Quantum Register
c = ClassicalRegister(1) #Initialize Classical Register for measurement results
circ = QuantumCircuit(q,c) #Create Quantum Circuit
circ.rx(pi/6,q) #30 degree rotation about X-axis
circ.draw('mpl')
```

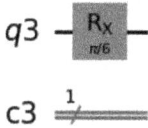

```
state = Statevector(circ)
print(state)
plot_bloch_multivector(state)
```

```
Statevector([0.96592583+0.j      , 0.        -0.25881905j],
            dims=(2,))
```

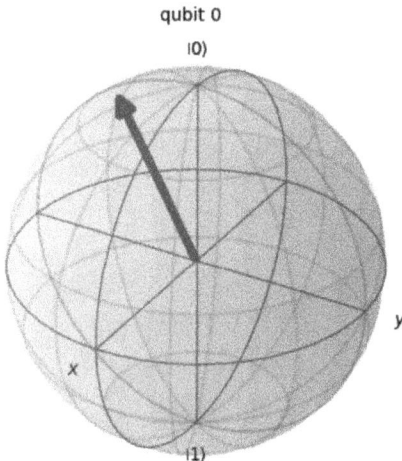

2. $\mathbf{R}_y(\theta)$ where $\theta = 30° = \frac{\pi}{6}$ rad:

This time the operation rotated the qubit from the default state of $|0\rangle$ on the xz-plane. Note that the $|0\rangle$ component is the same as in the previous example, but in this case the qubit has a component along the x-axis which is real because of the $\mathbf{R}_y$ matrix.

```
q = QuantumRegister(1) #Initialize Quantum Register
c = ClassicalRegister(1) #Initialize Classical Register for measurement results
circ = QuantumCircuit(q,c) #Create Quantum Circuit
circ.ry(pi/6,q) #30 degree rotation about Y-axis
circ.draw('mpl')
```

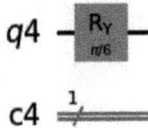

```
state = Statevector(circ)
print(state)
plot_bloch_multivector(state)
```

```
Statevector([0.96592583+0.j, 0.25881905+0.j],
            dims=(2,))
```

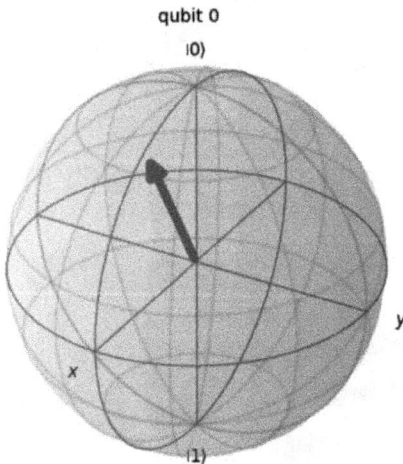

3. Finally, let us keep the $\mathbf{R}_y\left(\frac{\pi}{6}\right)$ rotation and add a $\mathbf{R}_z\left(\frac{\pi}{2}\right)$ rotation so that we will be able to observe the effect of the $\mathbf{R}_z$ gate:

As we can see, the $\mathbf{R}_z$ rotation further rotated our state from the xz-plane back to the yz-plane.

```
q = QuantumRegister(1) #Initialize Quantum Register
c = ClassicalRegister(1) #Initialize Classical Register for measurement results
circ = QuantumCircuit(q,c) #Create Quantum Circuit
circ.ry(pi/6,q) #30 degree rotation about Y-axis
circ.rz(pi/2,q) #90 degree rotation about Z-axis
circ.draw('mpl')
```

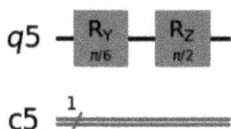

```
state = Statevector(circ)
print(state)
plot_bloch_multivector(state)
```

```
Statevector([0.6830127-0.6830127j, 0.1830127+0.1830127j],
            dims=(2,))
```

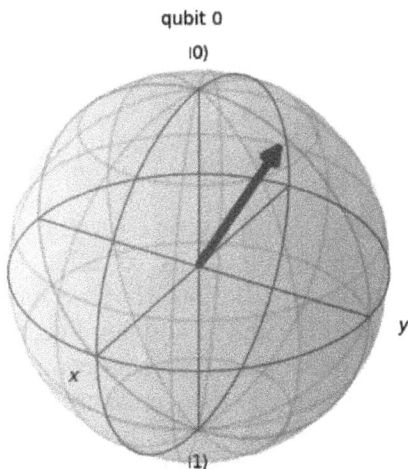

4.2.3   Arbitrary Rotation Gate

A general rotation, about any axis, $\hat{n}$, takes the state

$$|\psi\rangle = \cos\left(\frac{\theta}{2}\right)|0\rangle + \sin\left(\frac{\theta}{2}\right)e^{i\varphi}|1\rangle$$

which provided a clear correspondence between the qubit and the surface of the unit sphere and perform the following transformation:

$$|\psi'\rangle = U|\psi\rangle$$

If we tried to transform the $|0\rangle$ state using general unitary, U, we know it should take $|0\rangle \rightarrow |\psi\rangle$. This gives us the first column of the unitary matrix. Then, the second column will tell us what happens to the $|1\rangle$ state, since we are working in the $\{|0\rangle, |1\rangle\}$ basis. As we have learned, the second column (in general) will consist of two complex numbers, a and b, which have four unknowns:

$$U = \begin{pmatrix} \cos(\theta/2) & a \\ e^{i\varphi}\sin(\theta/2) & b \end{pmatrix}$$

However, we also know that the unitary matrix must satisfy the constraint: $U^\dagger U = I$. Writing this down, we have

$$U^\dagger U = \begin{pmatrix} \cos(\theta/2) & e^{-i\varphi}\sin(\theta/2) \\ a^* & b^* \end{pmatrix}\begin{pmatrix} \cos(\theta/2) & a \\ \sin(\theta/2)\,e^{i\varphi} & b \end{pmatrix} = \begin{pmatrix} 1 & 0 \\ 0 & 1 \end{pmatrix}$$

When we perform the matrix multiplication on the left-hand side, we obtain 3 equations, which allows us to solve for: $a = -e^{-i\lambda}\sin(\theta/2)$ and $b = e^{-i(\lambda+\varphi)}\cos(\theta/2)$. Note that we have introduced a new phase angle, λ. This is because a and b are general complex numbers, however, we saw previously that this complex phase (also known as global phase) does not affect the measurement results because for a general single qubit state $|\psi\rangle = \alpha|0\rangle + \beta|1\rangle$, we need to take $|\alpha|^2$ and $|\beta|^2$ which makes the effect of multiplying by any complex phase undetectable.

Finally, we can express the general unitary gate as

$$U(\theta, \varphi, \lambda) = \begin{pmatrix} \cos(\theta/2) & -e^{i\lambda}\sin(\theta/2) \\ e^{i\varphi}\sin(\theta/2) & e^{i(\lambda+\varphi)}\cos(\theta/2) \end{pmatrix}$$

We may express the Pauli gates and other single-qubit gates as specific cases of this general unitary gate. Alternatively, the Pauli gates also form a *basis*, $\{I, X, Y, Z\}$, with which any other single-qubit gate can be written. For example, you may check that the rotation matrices we introduced in the previous section can be equivalently expressed as follows:

$$R_x(\theta) = U(\theta, -\pi/2, \pi/2) = \cos(\theta/2)I - i\sin(\theta/2)X$$
$$R_y(\theta) = U(\theta, 0, 0) = \cos(\theta/2)I - i\sin(\theta/2)Y$$
$$R_z(\varphi) = U(0, 0, \varphi) = \cos(\varphi/2)I - i\sin(\varphi/2)Z$$

Additionally, the unitary matrix may be used to express *Clifford gates*. Clifford gates have the property of transforming one Pauli gate to another Pauli gate. An important Clifford gate is the Hadamard gate:

$$H = \frac{1}{\sqrt{2}}\begin{pmatrix} 1 & 1 \\ 1 & -1 \end{pmatrix} = U(\pi/2, 0, \pi)$$

The Hadamard gate is the same as performing $\mathbf{R}_y\left(\frac{\pi}{2}\right)$ operation. The Hadamard gate is important because it generates equal superposition states:

$$H\left|0\right\rangle = \frac{1}{\sqrt{2}}(\left|0\right\rangle + \left|1\right\rangle)$$

$$H\left|1\right\rangle = \frac{1}{\sqrt{2}}(\left|0\right\rangle - \left|1\right\rangle)$$

Let's implement the gate in Qiskit, starting with the default state $\left|0\right\rangle$:

```
q = QuantumRegister(1) #Initialize Quantum Register
c = ClassicalRegister(1) #Initialize Classical Register for measurement results
circ = QuantumCircuit(q,c) #Create Quantum Circuit
circ.h(0) #Hadamard gate
circ.draw('mpl')
```

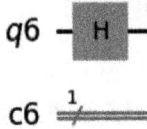

```
state = Statevector(circ)
print(state)
plot_bloch_multivector(state)
```

```
Statevector([0.70710678+0.j, 0.70710678+0.j],
            dims=(2,))
```

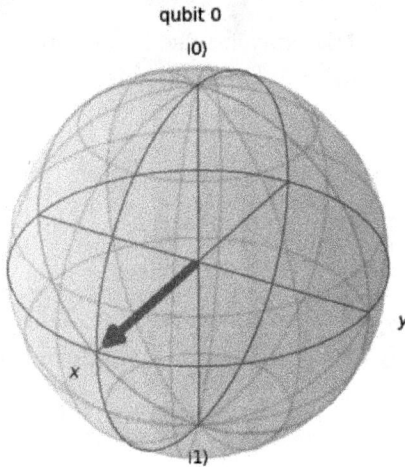

Now let's have the initial state as $|1\rangle$:

```
q = QuantumRegister(1) #Initialize Quantum Register
c = ClassicalRegister(1) #Initialize Classical Register for measurement results
circ = QuantumCircuit(q,c) #Create Quantum Circuit
circ.x(0) #Pauli-X gate
circ.h(0) #Hadamard gate
circ.draw('mpl')
```

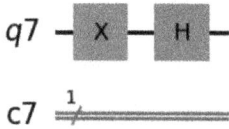

```
state = Statevector(circ)
print(state)
plot_bloch_multivector(state)
```

```
Statevector([ 0.70710678+0.j, -0.70710678+0.j],
            dims=(2,))
```

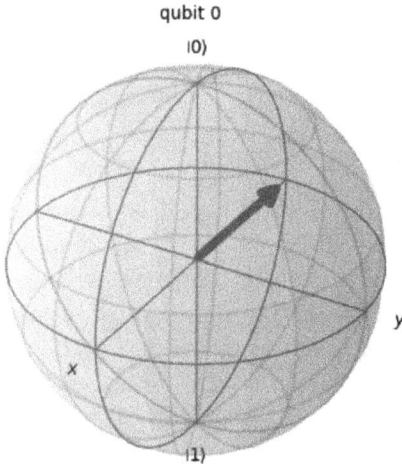

Checkpoint Exercises

1. Verify that the eigenvectors for the Pauli matrices are as given in Section 4.2.1.
2. Show that the condition $U^\dagger U = \mathbb{I}$ in 4.2.3 yields $a = -e^{-i\lambda}\sin(\theta/2)$ and $b = e^{-i(\lambda+\varphi)}\cos(\theta/2)$.

3. What rotation gate(s) can we use to generate the following state? Implement those gates and visualize this state in Qiskit:

$$|\psi\rangle = \frac{2}{\sqrt{3}}|0\rangle + \frac{1}{\sqrt{3}}|1\rangle$$

4. What gate(s) can act on the state $|1\rangle$ to produce the following superposition? Implement those gates and visualize this state in Qiskit:

$$|\psi\rangle = \frac{1}{\sqrt{2}}(|0\rangle - i|1\rangle)$$

4.3   Describing Measurement

In quantum mechanics, a state is represented by the probability amplitudes, c_1 and c_2, which may be complex numbers. In order to obtain probability, which has to be real and positive, we take the squared magnitudes of c_1 and c_2 to get $|c_1|^2$ and $|c_2|^2$. The normalization condition is therefore interpreted as probability conservation:

$$|c_1|^2 + |c_2|^2 = 1$$

When expressing an arbitrary qubit state $|\psi\rangle$, it is critical to specify a basis in which we can write the state. For example, if we are trying to measure the z-component of the spin, then we will express $|\psi\rangle$ in terms of the Z-basis which is our familiar $\{|0\rangle, |1\rangle\}$ set of basis vectors. If we were measuring the x-component of spin, then we would express $|\psi\rangle$ in terms of X-basis that are the states $\{|+\rangle, |-\rangle\}$, which may be written as a superposition of the Z-basis states:

$$|+\rangle = \frac{1}{\sqrt{2}}(|0\rangle + |1\rangle)$$

$$|-\rangle = \frac{1}{\sqrt{2}}(|0\rangle - |1\rangle)$$

Figure 4.3 provides a visual way of thinking about different basis states. Two orthogonal states (no overlap with each other) will serve as the axes with which we can specify a quantum state. States in different bases may be rewritten in terms of each other as necessary.

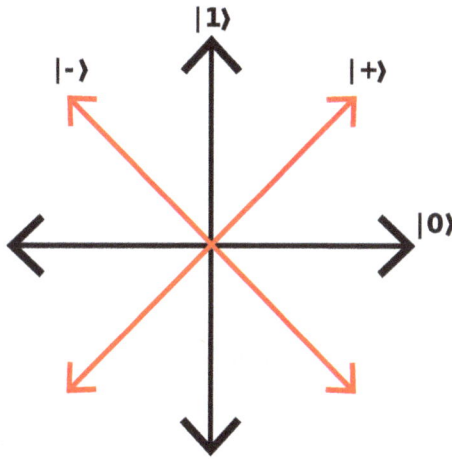

Figure 4.3. *X*-basis vs. *Z*-basis.

The way that we express the quantum state has a lot to do with what we are trying to measure. The basis that the measuring device works in is known as the **measurement basis**. Sometimes, we can have a measuring device oriented along the state of the qubit, and sometimes, the measurement device is fixed so it can only distinguish states that are oriented along its direction. Thus, we would need to rotate our qubit in some way to align with the basis of the measurement device. For spin, in this case, the measurement basis can be specified in terms of what component of spin we are interested in measuring (up, down,...). It is critical to have the state we want to measure align with the measurement basis and vice versa because if we measure in the wrong basis, we will force the quantum state to an incompatible basis and lose information. The act of measurement destroys the original superposition implying that after measurement we only have classical information.

Once we have expressed the state in the appropriate basis, then we can interpret the coefficients in front of the basis states as giving us the probability of getting the corresponding basis state. If we were trying to measure the z-component of spin, then we would express $|\psi\rangle$ in terms of the Z-basis, and we would have a probability of $|c_1|^2$ to get $|0\rangle$ and a probability of $|c_2|^2$ to get $|1\rangle$. We would obtain such statistics by preparing many states $|\psi\rangle$, running them through a SG apparatus oriented in the z direction, and keeping track of the number of times we get $|0\rangle$ or $|1\rangle$. With these statistics, we could construct a histogram to represent the probability distribution. Using this probability distribution, we can estimate the expected outcome from measuring the quantum state. The eigenvalues of the observable or gate that we are measuring represent the outcome of the measurement. For z-component of spin, the corresponding observable would be the Pauli Z matrix whose eigenvalues are -1 and 1.

4.3.1   Computing the Result of a Measurement

Now that we are familiar with measuring a quantum state, we will go through the mathematics that yields the measurement result. Recall the concept of the inner product that we discussed before. The inner product represents the projection of one vector onto another vector, or how much they overlap. For example, if we have the vector $|\psi\rangle$ and we wanted to know how much this vector overlaps with $|0\rangle$ or $|1\rangle$, we can take the inner product as follows:

$$\langle 0|\psi\rangle = c_1 \langle 0|0\rangle + c_1 \langle 0|1\rangle$$

But we said that $|0\rangle$ and $|1\rangle$ are orthogonal, so their overlap is 0 ($\langle 0|1\rangle = 0$) and a vector completely overlaps with itself ($\langle 0|0\rangle = 1$) so that inner product is 1. Therefore,

$$\langle 0|\psi\rangle = c_1$$

Similarly,

$$\langle 1|\psi\rangle = c_2$$

All we have shown here is another way of saying the components of $|\psi\rangle$ are c_1 along $|0\rangle$ and c_2 along $|1\rangle$. If we wanted to figure out the probability that we obtain the state $|0\rangle$ or $|1\rangle$,

$$\mathbb{P}(|0\rangle) = |\langle 0|\psi\rangle|^2 = |c_1|^2$$
$$\mathbb{P}(|1\rangle) = |\langle 1|\psi\rangle|^2 = |c_2|^2$$

Note, however, that if $|0\rangle$ and $|1\rangle$ were not orthogonal, we would not be able to trust the results of our measurements because there would be some overlap between the two states. This is why we will always choose a computational basis that is orthogonal and normalized or **orthonormal**.

Suppose we wanted to measure the x-component of spin of an arbitrary single qubit state, $|\psi\rangle = c_1 |0\rangle + c_2 |1\rangle$. The observable corresponding to this is the **X** matrix, so we can rewrite $|\psi\rangle$ in terms of the eigenvectors of **X**:

Then, we can express $|\psi\rangle$ in this new basis:

$$|\psi\rangle = d_1 |u_1\rangle + d_2 |u_2\rangle$$

Note here that $|0\rangle$ and $|1\rangle$ as superpositions of $|u_1\rangle = \frac{1}{\sqrt{2}} \binom{1}{1}$ and $|u_2\rangle = \frac{1}{\sqrt{2}} \binom{1}{-1}$:

$$|0\rangle = \frac{1}{\sqrt{2}} (|u_1\rangle + |u_2\rangle)$$

$$|1\rangle = \frac{1}{\sqrt{2}} (|u_1\rangle - |u_2\rangle)$$

Plugging into $|\psi\rangle = c_1 |0\rangle + c_2 |1\rangle$, we can solve for the coefficients d_1 and d_2 in terms of c_1 and c_2:

$$|\psi\rangle = \left(\frac{c_1 + c_1}{\sqrt{2}} \right) |u_1\rangle + \left(\frac{c_1 - c_2}{\sqrt{2}} \right) |u_2\rangle$$

If we measured this state, our apparatus would either read $+1$, which is the eigenvalue corresponding to $|u_1\rangle$ with probability $\left| \frac{c_1 + c_1}{\sqrt{2}} \right|^2$ or -1, the eigenvalue corresponding to $|u_2\rangle$ with probability $\left| \frac{c_1 - c_2}{\sqrt{2}} \right|^2$.

Knowing all this information, we may now compute the expected value of performing an **X**-gate on the state $|\psi\rangle = c_1 |0\rangle + c_2 |1\rangle$. Recall from our discussion of discrete random variables in Chapter 2. In this case, eigenvalues of the matrix can be considered as the random variables and the probabilities are part of the probability mass function of getting one particular eigenstate of the matrix. So, first, we rewrote $|\psi\rangle$ from $\{|0\rangle, |1\rangle\}$ basis in terms of the $\{|u_1\rangle, |u_2\rangle\}$ basis and obtained the coefficients d_1 and d_2, which gave us the probabilities of $|\psi\rangle$ being in either $|u_1\rangle$ or $|u_2\rangle$. In this case, the expected value for applying the X-gate can be computed as follows:

$$\langle\psi| X^\dagger X |\psi\rangle = \left|\frac{c_1 + c_1}{\sqrt{2}}\right|^2 - \left|\frac{c_1 - c_2}{\sqrt{2}}\right|^2$$

The term $\langle\psi| X^\dagger X |\psi\rangle$ is simply the inner product of the resulting state $\mathbf{X} |\psi\rangle$, and the left side is each probability scaled by its corresponding eigenvalue.

Let's do some simple checks for this result. If $|\psi\rangle = |0\rangle$, which has $c_1 = 1$ and $c_2 = 0$ which corresponds to the positive z-axis of the Bloch sphere or vice versa $|\psi\rangle = |1\rangle$, where $c_1 = 0$ and $c_2 = 1$, which corresponds to the negative z-axis of the Bloch sphere, then we would not expect any x-component for spin, so the above equation should equal 0. If $|\psi\rangle = |u_1\rangle$, then we expect to obtain $+1$, and if $|\psi\rangle = |u_2\rangle$, then we expect to obtain -1.

It is important to note that the Hadamard gate allows us to switch from the z-measurement basis, $\{|0\rangle, |1\rangle\}$, to the x-measurement basis, $\{|+\rangle, |-\rangle\}$, where $|+\rangle \Longleftrightarrow |u_1\rangle$ and $|-\rangle \Longleftrightarrow |u_2\rangle$. So, the Hadamard gate can be interpreted as a way to move information around a qubit since it is exchanging information that could be measured from the X-basis to the Z-basis and vice versa.

Remember that in order for us to build up a probability distribution, we need many trials. We have to prepare a large number of $|\psi\rangle$ and measure them to gather enough statistics. Then, we are able to calculate the average result that we would expect if we measured the state $|\psi\rangle$. If we just have one $|\psi\rangle$, then we only have "one trial," so the apparatus will yield $+1$ or -1 and the superposition will collapse to the corresponding eigenstate, $|u_1\rangle$ or $|u_2\rangle$, leaving us with limited knowledge about the original state. This is the same thing we saw with the light polarization example. Once the light passed through the polarizer, where only what got through would continue to propagate, the information about the polarization of the original source of light is lost.

4.3.2 Qiskit Example: Characterizing Single Qubit States with Measurement

Thus far, we have learned about how to rotate qubits around the Bloch sphere, studied the importance of measuring in the right basis, and learned how to compute expected values of some observable quantity like the spin in the x-direction. Now, let's use this knowledge to characterize quantum states. In practice, if we have some unknown quantum state, then there is no way for us to know the state vector. The only way we can interact with the state to learn about it is through measurement. It turns out that if it is possible to prepare many of the quantum states and perform measurements in different bases, then we can get enough information to fully characterize its state.

We will implement the solution to such a problem in Qiskit. The ground rules are that Qiskit has a fixed measurement basis $\{|0\rangle, |1\rangle\}$, so we will only be able to gain insight into the z-component of the state on the Bloch sphere. We can learn about the other axes if we perform additional rotations on the qubit so that we can measure other components. To perform measurements, we will be using the *qasm simulator* backend which provides us with measurement counts for the output of measuring a particular quantum state. We can specify the number of times we measure. In practice, this means that we prepare many identical quantum states and we measure each and record the output. The data are organized using a histogram which tells us what states were output and with what probability those states occurred.

Let's do an example with a state that someone else prepared a quantum state and we do not know what it is. Let's say the state is given by the following circuit:

```
q = QuantumRegister(1) #Initialize Quantum Register
c = ClassicalRegister(1) #Initialize Classical Register for measurement results
circ = QuantumCircuit(q,c) #Create Quantum Circuit
circ.ry(pi/4,q) #45 degree rotation about Y-axis
circ.draw('mpl')
```

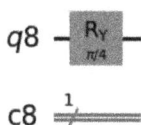

$q8$ — $\boxed{\begin{array}{c} R_Y \\ \pi/4 \end{array}}$ —

$c8$ ═╪═══

```
state = Statevector(circ)
print(state)
plot_bloch_multivector(state)
```

```
Statevector([0.92387953+0.j, 0.38268343+0.j],
            dims=(2,))
```

qubit 0

|0⟩

y

x

|1⟩

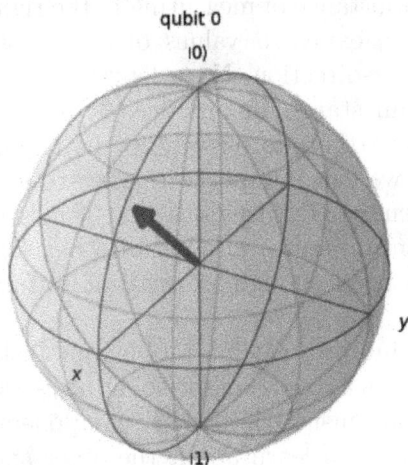

If we are given such a quantum state and asked to figure out its x-, y-, and z-components on the Bloch sphere, we need to perform thoughtful measurements of the qubit, knowing that the Qiskit measurement is fixed to only measure along the z-axis. But we can look at different components of the state is by performing additional rotations. Specifically, we want to be able to rotate the x- and y-axes to align with the default position of the z-axis. To move the current x-axis to the position of the current z-axis, we need to perform a rotation about the y-axis by 90° clockwise, as shown in Fig. 4.4. To move the current y-axis to the position of the current z-axis, we would need to rotate about x-axis by 90° counterclockwise, as shown in Fig. 4.5.

Then we can calculate the expected value of measuring along each of the axes and we would get the spin x-, y-, and z-components of the state, where the eigenvalues for each are still 1, −1 and the probabilities will be given from the histogram.

Measurement of the z-component

Let's start off by first measuring the z-component which does not require doing any rotation. We will use a function called "deepcopy"

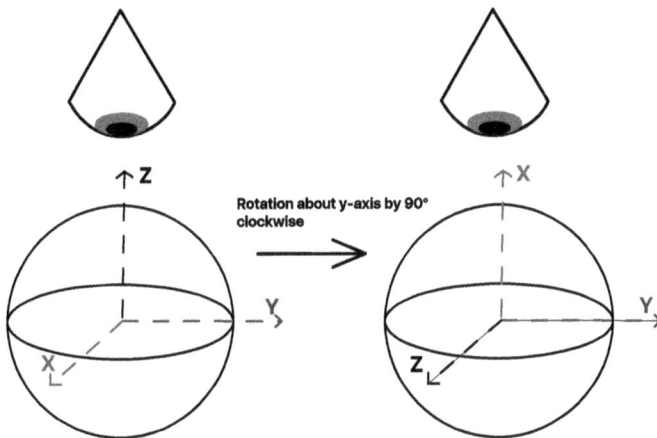

Figure 4.4. By performing a rotation of 90° clockwise about the y-axis, the z- and x-axes swap so that the measuring device measures along x.

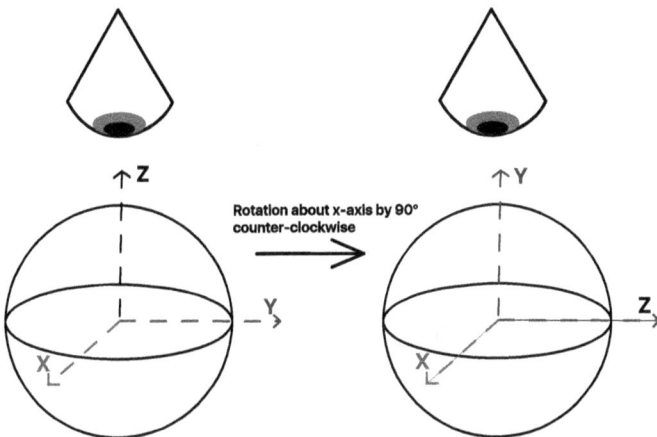

Figure 4.5. By performing a rotation of 90° counterclockwise about the x-axis, the z- and y-axes swap so that the measuring device measures along y.

which essentially copies the Quantum Circuit object in Python. We may physically interpret this action as preparing another state identical to the original so that we may just focus on measuring the z-component. Physically, this can be done with physical quantum systems that can be initialized through their ground states. Continuing on from the previous code, we can write the following code, where we are relabeling the "copied" circuit as "circZ" denoting that we want to measure the z-component. In the "qasm simulator," we specify 100 shots, meaning we prepared 100 states that are identical, and we measure each of them and record the output. Then, with the probabilities we get, we can calculate the expected value for measuring along the z-axis as follows, remembering that our random variables are -1 and 1.

$$\mathbb{E}z = \mathbb{P}(0) - \mathbb{P}(1)$$

```
from copy import deepcopy
circZ = deepcopy(circ) #Copy original Quantum Circuit object
circZ.measure(q,c)
circZ.draw('mpl')
```

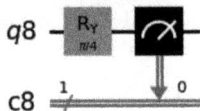

```
shots = 100
simulator = Aer.get_backend('qasm_simulator')
circ_transpile = transpile(circZ, backend = simulator)
result = simulator.run(circ_transpile,shots = shots).result()
counts = result.get_counts()
print(counts)
plot_histogram(counts)
```

`{'1': 11, '0': 89}`

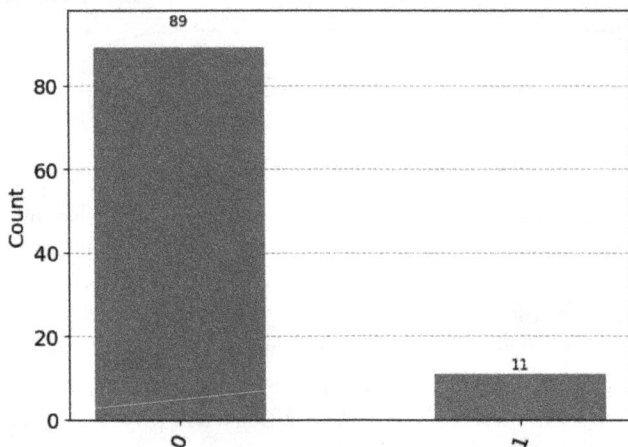

Therefore, using the equation from above, the z-component of this qubit would be $0.89 - 0.11 = 0.78$.

Measuring the y-component

Now, let us repeat the same procedure by adding the necessary rotation to swap the z- and y-axes through the 90° counterclockwise rotation about the x-axis. The expected value of measuring along the y-axis is still

$$\mathbb{E}y = \mathbb{P}(0) - \mathbb{P}(1)$$

```
circY = deepcopy(circ) #Copy original Quantum Circuit object
circY.rx(pi/2,0) #90 degree rotation about X-axis to swap Z and Y
circY.measure(q,c)
circY.draw('mpl')
```

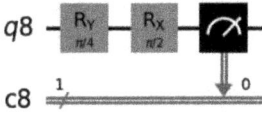

```
shots = 100
simulator = Aer.get_backend('qasm_simulator')
circ_transpile = transpile(circY, backend = simulator)
result = simulator.run(circ_transpile,shots = shots).result()
counts = result.get_counts()
print(counts)
plot_histogram(counts)
```

`{'1': 52, '0': 48}`

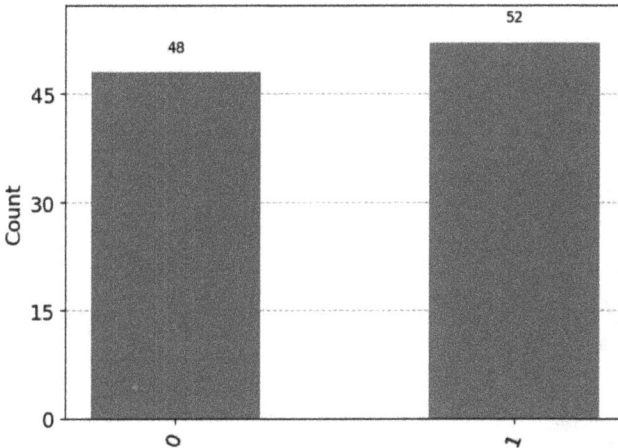

So, the y-component for the qubit is $0.48 - 0.52 = -0.04$. Glancing back at the original state on the Bloch sphere, the state is positioned on the xz-plane, so it makes sense that we do not really have any

y-component. The reason why we still computed some y-component is due to counting errors since we only have 100 counts.

Measuring the x-component

Finally, let us repeat the same procedure by adding the necessary rotation to swap the z- and x-axes through the 90° clockwise rotation about the y-axis. The expected value of measuring along the x-axis is still

$$\mathbb{E}x = \mathbb{P}(0) - \mathbb{P}(1)$$

So, the x-component for the qubit is $0.81 - 0.19 = 0.62$

```
circX = deepcopy(circ) #Copy original Quantum Circuit object
circX.ry(-pi/2,0) #-90 degree rotation about Y-axis to swap Z and X
circX.measure(q,c)
circX.draw('mpl')
```

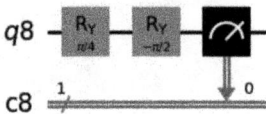

```
shots = 100
simulator = Aer.get_backend('qasm_simulator')
circ_transpile = transpile(circX, backend = simulator)
result = simulator.run(circ_transpile,shots = shots).result()
counts = result.get_counts()
print(counts)
plot_histogram(counts)
```

`{'1': 19, '0': 81}`

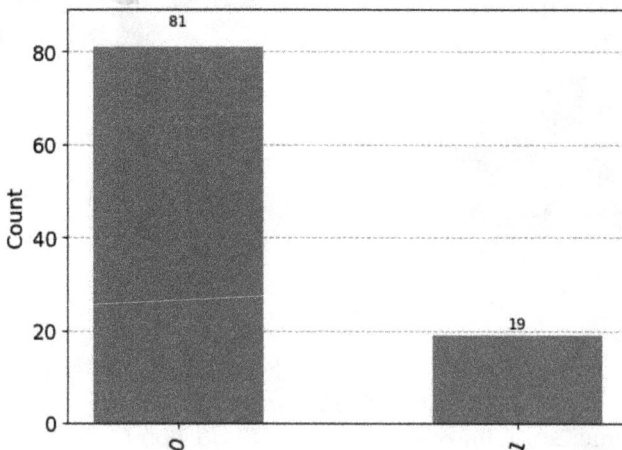

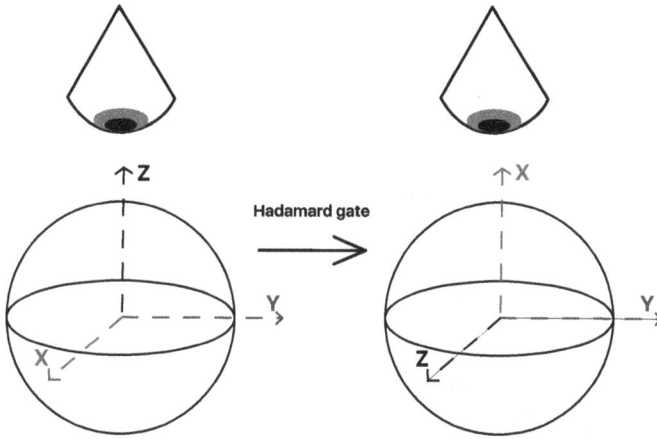

Figure 4.6. With the Hadammard gate, the $+x$-axis is in the place of $|1\rangle$, but it represents $|0\rangle$, while the $-x$-axis is in the place of $|0\rangle$, but it represents $|1\rangle$. So, we have to swap the eigenvalues that correspond to these states.

It is interesting to note that instead of the $90°$ clockwise rotation about the y-axis, we may also use the Hadamard gate which also lets us swap the z- and x-axes, as shown in Fig. 4.6. The expected value is still

$$\mathbb{E}x = \mathbb{P}(0) - \mathbb{P}(1)$$

We can implement this in Qiskit in a similar fashion. So, the x-component for the qubit is $0.85 - 0.15 = 0.7$, which is close to what we got doing the y-rotation. Of course, the results are probabilistic, so they will be different.

```
circX = deepcopy(circ) #Copy original Quantum Circuit object
circX.h(0) #-90 degree rotation about Y-axis to swap Z and X
circX.measure(q,c)
circX.draw('mpl')
```

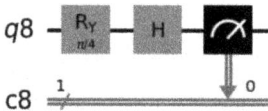

```
shots = 100
simulator = Aer.get_backend('qasm_simulator')
circ_transpile = transpile(circX, backend = simulator)
result = simulator.run(circ_transpile,shots = shots).result()
counts = result.get_counts()
print(counts)
plot_histogram(counts)
```

{'1': 15, '0': 85}

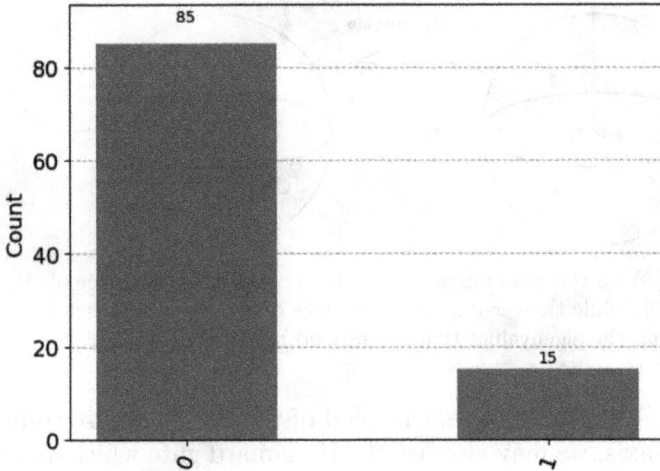

Checkpoint Exercise

Your friend gives you a quantum state that can be represented by the following circuit:

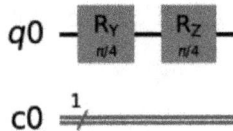

1. Implement the circuit in Qiskit and plot the quantum state on the Bloch sphere to visualize it.
2. Now, pretend you do not know what the state is and perform measurements to figure out the x-, y-, and z-components of the state. Use 100 shots for running the "qasm simulator."
3. Verify your results with the Bloch vector you plotted in number 1.

4. Compute the magnitude of the vector:

$$\sqrt{x^2 + y^2 + z^2}$$

   Does it equal 1? Why or why not?
5. If you increase the number of shots to 1000, what is the magnitude of the vector now? Is it closer to 1?

4.3.3 Projective Measurements of Observables

As we have seen thus far, measurement of qubits could be tricky because we have to ensure that we are in the right basis so that our measurement outcomes make sense. Suppose we are trying to measure some particular observable, M, which is Hermitian because $M^\dagger = M$. This operator will have real eigenvalues, λ_j, and an orthonormal set of eigenvectors which we will denote as $\{|\phi_j\rangle\}$. So, M can be expressed as follows[2]:

$$M = \sum_{j=1}^{N} \lambda_j |\phi_j\rangle \langle\phi_j|$$

So now, if we have an arbitrary qubit, $|\psi\rangle$, we can write the state in the eigenbasis of the observable we want to measure:

$$|\psi\rangle = c_1 |\phi_1\rangle + \cdots + c_N |\phi_N\rangle$$

Now, depending on the observable we have, the results we obtain from the measurement apparatus will be one of the eigenvalues, λ_j, and these values must clearly be real so that we may measure them. The probability that we read off this eigenvalue is $|c_j|^2$. The state of $|\psi\rangle$ after measuring will be the eigenvector, $|\phi_j\rangle$ that corresponds to the eigenvalue, λ_j.

---

[2]The notation, $|u\rangle \langle u|$, represents an outer product which results in a matrix, for example,

$$|0\rangle \langle 0| = \begin{bmatrix} 1 \\ 0 \end{bmatrix} \begin{bmatrix} 1 & 0 \end{bmatrix} = \begin{bmatrix} 1*1 & 1*0 \\ 0*1 & 0*0 \end{bmatrix} = \begin{bmatrix} 1 & 0 \\ 0 & 0 \end{bmatrix}$$

4.3.4   No-Cloning Theorem

Our discussion of measurement gives rise to an important idea: since the act of measurement is destructive, causing the collapse of the superposition state, it would be impossible to clone or create an independent and identical copy of some unknown, arbitrary quantum state. In order to copy a state, we would need to first *know* what that state is, but if we *know* the state, the superposition will have collapsed, so we are unable to copy the original state.

This seems strange and limiting in some sense. For example, in classical computing, the majority of error-correcting codes involve using repetition or having backup copies of bits in case something went wrong. In quantum computing, *this* is clearly impossible! However, there is a bright side to this phenomena: highly secure communication with quantum which we will discuss more of in the following chapter.

Homework 4

1. Using the procedure we discussed for performing projective measurements in Qiskit, implement a function that takes in the following inputs:

   1. Quantum Register object,
   2. Classical Register object,
   3. Quantum Circuit object,
   4. Number of shots to run the qasm simulator.

   The output of the function should be an array or list object that contains the x-, y-, z-coordinates of the state obtained using only measurements from the qasm simulator.

   The necessary imports to run your code are as follows:

   ```
   #Import necessary packages to run code
   import numpy as np
   from math import pi
   from qiskit import *
   from qiskit.quantum_info import Statevector
   from qiskit.visualization import plot_bloch_
   multivector, plot_histogram
   from copy import deepcopy
   ```

You may take advantage of the following helper functions:

```
def getCounts(circuit, shots):
    """
    Inputs: Quantum Circuit and number of shots
    Returns: dictionary of measurement counts
    """
    simulator = Aer.get_backend('qasm_simulator')
    circ_transpile = transpile(circuit, backend =
    simulator)
    result = simulator.run(circ_transpile,shots =
    shots).result()
    counts = result.get_counts()

return result.get_counts(circuit)

def getExpectationValue(counts, shots):
    """
    Inputs: Dictionary of measurement counts and
    number of shots
    Returns: Expectation value
    """
    E = (counts.get('0',0) - counts.get('1',0))/
    shots
return E
```

2. Test your functions on the two example circuits shown in Figs. 4.7 and 4.8.

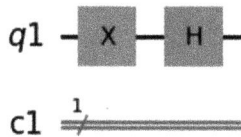

Figure 4.7. Circuit 1.

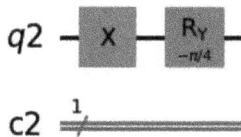

Figure 4.8. Circuit 2.

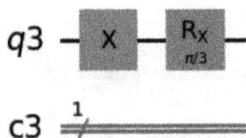

Figure 4.9.   Circuit 3.

Do some additional post-processing to convert the coordinates to angles using the following:

$$\tan \varphi = \frac{y}{x}$$

$$\cos \theta = z$$

Verify that if you plug these angles back into the Bloch State we derived before, its close to the state vector output by the "Statevector simulator":

$$|\psi\rangle = \cos\left(\frac{\theta}{2}\right)|0\rangle + \sin\left(\frac{\theta}{2}\right)e^{i\varphi}|1\rangle = \begin{pmatrix} \cos(\theta/2) \\ \sin(\theta/2)e^{i\varphi} \end{pmatrix}$$

Now, do one more case that is shown in Fig. 4.9.

You will find that there is a global phase inconsistency between the method we are using and Statevector. Is this is a concern in terms of measuring such a state?

Chapter 5

Applications with Single Qubits

5.1 Classical Cryptography

Cryptography is the art of concealing information from unauthorized third parties, while two authorized parties exchange information. This notion has been around since ancient times and continues to be used in the modern world for data protection in commercial and public institutions. Binary strings or numbers that are randomly chosen from a large set are known as **keys** and they provide the security for cryptographic protocols, including encryption, authentication and sharing.

There are two classes of keys: symmetric or public–private key pairs. Symmetric keys consist of one key or a pair of keys that can be easily computed from one to the other, and are only known to the communicating parties. In this case, the involved parties must all guard the key! Public–private key pairs comprise a public key which is known to all and a private key which has to be guarded by the owner.

The essential structure of a cryptographic scheme using a symmetric key is represented in Fig. 5.1. Before sending any secret information, Bob (the sender) sends a plain text with a secret key to which an encryption algorithm is applied to scramble the message, known as the cipher text. The cipher text is unscrambled using a decryption algorithm for Alice to read. The only way to understand the plain text from the cipher text is by knowing the key, so some eavesdropper, Eve, will be unable to deduce the secret message. In this case,

Figure 5.1.   Essentials of a cryptographic scheme using a symmetric key.

Figure 5.2.   Essentials of a cryptographic scheme using public–private key pairs.

the shared key is the same for both encryption and decryption.

The protocol structure of a cryptographic scheme using a public–private key pair is presented in Fig. 5.2. In this case, the public key is used to encrypt Bob's private message which can be decrypted by Alice using the matching private key. Alice is the private key owner and the one associated with the public key to maintain confidentiality.

It is fundamental that Alice and Bob establish keys of truly random numbers, so that an unauthorized third party, Eve, cannot deduce the key. Secondly, it is critical that Eve cannot *intercept* the secret throughout the exchange between Alice and Bob. One extra step to the process is the distribution of an open padlock in which Bob (the sender) has to authenticate this padlock to verify that it has not been tampered with and that it comes from the intended recipient, Alice. These padlocks are called "one-way" functions which are mathematical expressions that are easy to compute but very difficult to reverse. The most common example is the RSA public-key system which relies on factorization of prime numbers. So, two prime numbers are very easy to multiply but if only the product is known, it is very difficult to go backwards and find the two prime numbers that were multiplied together.

5.2 Quantum Key Distribution

Quantum key distribution protocols could be one of the first applications of quantum information processing that exploits quantum effects for security. The famous example we will discuss here is known as the BB84 protocol that was first invented in 1984 by IBM researcher Charles Bennett and Gilles Brassard from the University of Montreal [27]. The goal of this protocol is to establish a secret key that is known only to Alice and Bob. Alice and Bob want to use this key to exchange secret messages and detect potential tampering by Eve.

Consider that there are two channels of communication, a quantum channel and a classical channel which can both be observed by Eve, as shown in Fig. 5.3.

Firstly, over the quantum channel, Alice starts out by choosing a random sequence of bits which can be generated using a random number generator (e.g. 01101001). She also randomly chooses a basis to encode each bit in the sequence. In this situation, Alice transmits a photon for each bit where the polarization depends on the basis. As we discussed before in the photon polarization example, the choice of basis will determine our measurement results! In this case, we stick with the following linear polarization bases:

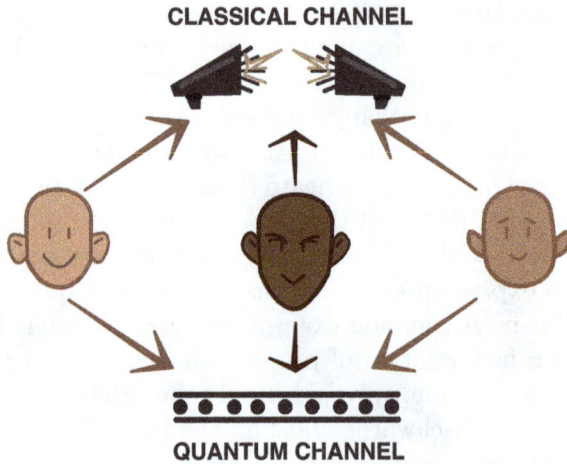

Figure 5.3. Quantum key distribution model and channels.

1. Standard basis (S):

$$0 \rightarrow |\uparrow\rangle$$
$$1 \rightarrow |\rightarrow\rangle$$

2. Diagonal basis (D):

$$0 \rightarrow |\nearrow\rangle$$
$$1 \rightarrow |\nwarrow\rangle$$

When Bob receives the photons, he also randomly selects either basis to measure the photons. Then, Alice and Bob use the classical channel to (1) check that Bob has indeed received each photon Alice sent and (2) compare the bases that they used to encode and measure the photons, respectively. If they both used the same basis, Bob's measured state agrees with the one Alice sent and if they didn't, there is only a 50% chance that Bob's result matches what Alice sent. Without actually revealing the bit values, they discard the ones for which the choice of basis was different. On average, 50% of Alice's originally transmitted bits remain part of the shared key, and depending on the level of security Alice and Bob want, they may compare some of the bit values to identify if eavesdropping occurred and discard the bits that do not match.

Table 5.1. Generating a private shared key between Alice and Bob using quantum key distribution.

Alice's random bit sequence	0	0	1	1	0	1	1	0
Alice's basis	S	S	S	D	D	S	D	S
Alice's polarization	$\lvert\uparrow\rangle$	$\lvert\uparrow\rangle$	$\lvert\rightarrow\rangle$	$\lvert\nwarrow\rangle$	$\lvert\nearrow\rangle$	$\lvert\rightarrow\rangle$	$\lvert\nwarrow\rangle$	$\lvert\uparrow\rangle$
Bob's basis	S	D	D	S	D	S	D	D
Bob's measurements	$\lvert\uparrow\rangle$	$\lvert\nearrow\rangle$	$\lvert\nearrow\rangle$	$\lvert\uparrow\rangle$	$\lvert\nearrow\rangle$	$\lvert\rightarrow\rangle$	$\lvert\nwarrow\rangle$	$\lvert\nearrow\rangle$
Private shared key	0				0	1	1	

Table 5.1 shows an example of such a procedure.[1]

So, how can Alice and Bob identify interceptions from Eve? Well, remember that over the classical channel, Alice and Bob only discuss what bases they chose but NOT the bit values themselves, so Eve cannot learn the key from listening to the classical channel. However, she can try to intercept the photons transmitted by Alice through the quantum channel. To do so, Eve has to send photons to Bob before knowing what their choice of bases were since bases confirmation between Alice and Bob occurs *after* Bob receives the photons. However, if Eve sends different photons to Bob, then Alice and Bob can detect the issue when they compare notes over the classical channel, while if she were to send the original photons to Bob, she would not gain any information.

Eve can also try to measure the photons Alice is sending before they get to Bob, but just like when Bob would measure the photons, there is 50% chance that Eve will measure in the wrong basis and thus cause a collapse of the initial states that Alice sent. Now, the photons

---

[1]Note that when the bases do not match, there is only a 50% chance that what Alice transmitted will be measured by Bob, for example,

$$\lvert\nearrow\rangle = \frac{1}{\sqrt{2}}(\lvert\uparrow\rangle + \lvert\rightarrow\rangle)$$

In this case, when Bob is in the diagonal basis and tries to measure a photon whose polarization is in the standard basis which results in only a 50% chance of measuring correctly $\left(\lvert\frac{1}{\sqrt{2}}\rvert^2 = \frac{1}{2}\right)$!

Bob receives have been tampered with, so even if Bob measures in the same basis as Alice, he can only measure the correct polarization states 50% of the time. Of the bits that remain, after Alice and Bob share what basis they used, there is a 25% chance that Bob measured a different bit value than what Alice originally sent.[2] This will be detectable by Alice and Bob when they compare a subset of the bits over the classical channel. If the bits agree, then they can use them as the private key, otherwise they know that Eve has intercepted.

Eve's problem is that she does not know what basis to measure the photons in before Bob receives them. If she could know the basis, then her interception would go unnoticed. The right way for her is to try to copy the state of the photons and send the original to Bob. Then, after learning the bases used while listening to the classical channel, she can measure her copies. This is impossible, however, because quantum states cannot be copied due to the no-cloning theorem! To copy something, it has to measured, and if a quantum state is measured, one cannot reliably *know* what the original state was unless the basis is known beforehand.

5.3   Quantum Bomb Detection

Physicists Avshalom C. Elitzur and Lev Vaidman published a paper in 1993 concerned with *interaction-free* measurements [28]. This principle was used to determine the existence of an object in some region of space without directly interacting with it through the use of quantum experiments. An intriguing example they proposed is to check a bomb without exploding it. Suppose that we are in a quantum factory and have the task of testing a newly developed light-sensitive bombs which have a special sensor installed. If a single photon hits the sensor, the bomb will explode. If the sensor is defective, then we consider the bomb to be a dud. However, we don't want to destroy

---

[2]This is the same concept as we discussed in the light polarization example. If the light wave coming in at a 45° angle, $|\nearrow\rangle$, passed through a polarizer positioned $|\uparrow\rangle$, then we would get 50% of the original wave amplitude. If we pass this wave through a $|\nearrow\rangle$ again, then it will only be 25% of the original amplitude! This is because of that mismatch in bases occurring in the first measurement.

Figure 5.4. Setup of quantum bomb problem.

the entire factory, so we need a way of testing whether the bombs are defective or not that guarantees with high probability that the bomb will not explode.

5.3.1 Classical Approach

We can think of the bomb as a black box since we do not know whether it works or not. Let's consider a simple probabilistic approach to test the bombs: we flip a coin and if heads (or $|0\rangle$) comes up, we declare the bomb to be live without testing. If tails (or $|1\rangle$) comes up, then we send a photon through and report the bomb as a dud if it does not explode. With such an approach, half the time, we will flip heads and can only correctly declare the bomb as not defective only half of those times. Similarly, if tails are flipped, the bomb will explode half of the time, but we can only be correct in calling it a dud half of that time. So, our success rate is 50% which is as bad as if we just tested each bomb one by one.

5.3.2 Quantum Setup

Now, perhaps using our knowledge of qubits, we can test whether the bomb works without actually exploding it. To do so, we exploit quantum measurement and quantum gates to increase our odds.

In the quantum setup of this problem, the bomb acts as a measurement device for the incoming photons or qubits. If the bomb is live, it performs a measurement on the qubit, and if the bomb is a dud, it does not. By developing a quantum circuit that implements some of the single-qubit gates we were introduced to in the previous sections, we can increase our odds of figuring whether the bomb is live without exploding it [29].

Let's consider the case that the incoming qubit (or photon) is in a superposition:

$$|\psi\rangle = \alpha |0\rangle + \beta |1\rangle$$

If the bomb happens to be a dud, then it does not perform a measurement, so the output state of the qubit remains $|\psi\rangle$. However, if the bomb is live, then the bomb measures the incoming qubit in the $\{|0\rangle , |1\rangle\}$ basis. The result will be $|0\rangle$ with probability $|\alpha|^2$, so the bomb will not explode even though it is live. On the other hand, there is $|\beta|^2$ probability, the outcome is $|1\rangle$ and the bomb will explode. So, we must do something to increase $|\alpha|^2$ and thus lower $|\beta|^2$.

First, we start with our qubit/photon in the $|0\rangle$ state and apply some rotation to the qubit to put in superposition using the gate $U(\theta, 0, 0)$, which we rename $R(\theta)$:

$$R(\theta) = \begin{pmatrix} \cos(\theta/2) & -\sin(\theta/2) \\ \sin(\theta/2) & \cos(\theta/2) \end{pmatrix}$$

Thus, we have the qubit going through the bomb:

$$R(\theta) |0\rangle = \cos(\theta/2) |0\rangle + \sin(\theta/2) |1\rangle$$

If $\theta = \frac{\pi}{2}$, then we would get the rotation gate:

$$R\left(\frac{\pi}{2}\right) = \frac{1}{\sqrt{2}} \begin{pmatrix} 1 & -1 \\ 1 & 1 \end{pmatrix}$$

Thus, the qubit state would be $|+\rangle = \frac{1}{\sqrt{2}}(|0\rangle + |1\rangle)$. So, if we sent $|+\rangle$ through a defective bomb, we would just get $|+\rangle$ back. If the bomb was live, it would perform a measurement and the output state would be $|0\rangle$ 50% of the time that the bomb has not exploded. If we ran the output of the bomb back through the same rotation gate, then we need to change the basis to the $\{|+\rangle , |-\rangle\}$ basis, so we can appropriately measure the results of the output.[3] If the bomb

---

[3]Note here that we want to distinguish $|+\rangle$ and $|0\rangle$, but their inner product (you may verify) $\langle +|0\rangle = \frac{1}{\sqrt{2}} \neq 0$, and if because of the live bomb performs a measurement, then we only obtain $|0\rangle$ 25% of the time. But, we get the state $|+\rangle$ 50% time when the bomb is a dud. So, we are exploiting this imbalance to ensure we measure the $|+\rangle$ state correctly.

was defective, we will obtain

$$R\left(\frac{\pi}{2}\right)|+\rangle = |1\rangle$$

Otherwise, if the bomb was not defective, we obtain $|0\rangle$ 50% of the time,

$$R\left(\frac{\pi}{2}\right)|0\rangle = |+\rangle$$

Now, there is a 50% chance from this result that we obtain either $|0\rangle$ or $|1\rangle$.

So, from the beginning, if the bomb was live, we had 50% chance of getting $|0\rangle$ after it was passed through the bomb. Then, we rotated again, so there was only a 50% chance there to get $|0\rangle$ at the output. In the last measurement, we only had a 25% chance to measure $|0\rangle$. Our overall success rate is therefore[4]:

$$\frac{1}{2} + \frac{1}{2} \cdot \frac{1}{4} = 62.5\%$$

There is a 62.5% chance of identifying the bomb without exploding! Can we do better?

Let's try to rotate the qubit by the angle, $\theta = \frac{\pi}{3}$, and send it through the bomb twice. So, our rotation gate is

$$R\left(\frac{\pi}{3}\right) = \begin{pmatrix} \frac{\sqrt{3}}{2} & -\frac{1}{2} \\ \frac{1}{2} & \frac{\sqrt{3}}{2} \end{pmatrix}$$

So, the input state will be

$$R\left(\frac{\pi}{3}\right)|0\rangle = \frac{\sqrt{3}}{2}|0\rangle + \frac{1}{2}|1\rangle$$

If the bomb is a dud, and we are sending this state through the bomb and rotating it two more times, at the end, we will measure and get $|1\rangle$ as expected because we have just rotated the qubit by $\frac{\pi}{2}$

---

[4]For any given bomb, 50% of the time, the bomb can be a dud and we always can correctly identify the dud. The other 50% of the time, the bomb is live, *but* we are able to say it's not a dud by increasing our chances of measuring $|0\rangle$ at the end. This computation of the success probability, $\mathbb{P}(\text{success}) = \mathbb{P}(D) + \mathbb{P}(UL)$, where D stands for dud, and UL stands for live bomb that were not exploded.

in total. If the bomb is live, the input state goes into the bomb once and has a $\frac{3}{4}$ chance of outputting $|0\rangle$ and a $\frac{1}{4}$ chance of getting $|1\rangle$ and detonating. Then, we run the $|0\rangle$ state through another $R\left(\frac{\pi}{3}\right)$ which goes through the bomb again. Now, the probability of detonating (getting $|1\rangle$) is reduced to $\frac{3}{4} \cdot \frac{1}{4} = \frac{3}{16}$, and thus, $\frac{9}{16}$ of the time, the output is $|0\rangle$ and it makes it through. Finally, it is rotated again to ensure we can distinguish the original state from $|0\rangle$. So, at the end, we measure $|0\rangle$, $\frac{9}{16} \cdot \frac{3}{4} = \frac{27}{64}$. Our probability of success is therefore

$$\frac{1}{2} + \frac{1}{2} \cdot \frac{27}{64} = \frac{91}{128} \approx 71\%$$

5.3.3   Generalization

Thus far, we saw that if we performed a $\frac{\pi}{2}$ rotation, we needed to repeat the rotation twice, for a $\frac{\pi}{3}$ rotation, we repeated the rotation three times. If we perform a rotation of angle $\epsilon = \frac{\pi}{N}$, where N is the number of repetitions by which we repeat the process of rotating the $|0\rangle$ qubit and sending it through the bomb, then the probability of getting $|1\rangle$ and the bomb detonating is[5]

$$\sin^2\left(\frac{\epsilon}{2}\right) \approx \frac{\epsilon^2}{4}$$

The total probability of setting off the bomb is

$$N\sin^2\left(\frac{\epsilon}{2}\right) \approx N\frac{\epsilon^2}{4} = \frac{\pi^2}{4N}$$

Thus, the probability of success is

$$\mathbb{P}(\text{success}) = 1 - \frac{\pi^2}{4N}$$

So, if we were to repeat the process $N = 100$ times, then we would have $\approx 97.5\%$ success rate of testing bombs and not exploding them. Now, we have devised a safe way to test these bombs and not burn down our quantum factory.

---

[5]As N gets really large, the angle ϵ gets very small, which allows us to perform a Taylor expansion to the sine function for small ϵ. For small angles, $\sin\theta \approx \theta$, so $\sin^2\theta \approx \theta^2$.

Homework 5

1. Alice and Bob are teachers that want to come up with exam questions and they want to communicate securely so that their very tech-savvy student Eve does not learn the questions prior to the exam. In order to communicate securely, they must establish a private secret key because they know that Eve will try to eavesdrop on their conversation. They decide it's best to deploy the BB84 protocol, so they can detect whether Eve eavesdropped. They use a quantum channel to send qubits and each randomly chooses a basis to encode the key. They communicate over the classical channel each time Alice sends the qubits, so she can verify that Bob got them.

   Suppose that the standard basis means Z-basis corresponding to the polarization states $|0\rangle$, $|1\rangle$ and diagonal basis means X-basis corresponding the polarization states, $|+\rangle$, $|-\rangle$. Let's also suppose that Alice has chosen the following random bit sequence, basis, and Bob has chosen the following basis and performed the measurements:

   Table 5.2. Generating a private shared key between Alice and Bob using quantum key distribution.

Alice's random bit Sequence	0	0	1	1	0	1	1	0
Alice's basis	Z	X	Z	X	X	Z	X	Z
Alice's polarization	$|0\rangle$	$|+\rangle$	$|1\rangle$	$|-\rangle$	$|+\rangle$	$|1\rangle$	$|-\rangle$	$|0\rangle$
Bob's basis	Z	Z	X	Z	X	Z	X	X
Bob's measurements	$|0\rangle$	$|0\rangle$	$|+\rangle$	$|0\rangle$	$|+\rangle$	$|0\rangle$	$|-\rangle$	$|+\rangle$

   Based on the table, what should be the privately shared key? Has Eve eavesdropped? How do you know?

Followup Questions

a. Assuming Eve is sleeping and cannot eavesdrop, how many bits should Alice randomly generate in order to use the BB84 protocol if she wants to have a bit-key string that is 256 bits? (*Hint:* Consider what is the probability of Bob choosing the same basis as Alice when he measures.)

b. Now, Eve is awake. Alice and Bob compare the measurement bases at the end of their transfer of bits but do not share information about the bits themselves. What is the probability that Eve can correctly guess the bit for a **single bit key**?

c. If Alice sends Bob 40 keys and they perform the protocol ending up with a 20 bit key in which their bases matched, what is the probability that Eve could have eavesdropped on all 20 bits without being detected?

2. We have discussed how to resolve the Elitzur Vaidmann Bomb problem. Now, let's try to implement the problem in Qiskit. The idea is if we have a bomb that is live, we want to rotate it by the small angle ε and measure so that it can keep collapsing to the $|0\rangle$ state and not detonate. If the bomb is not live, we rotate and measure just once and get $|1\rangle$, but the bomb will not explode because it is a dud. You may copy and use the following code as a skeleton to set up your solution. The green text are Python comments to help guide you through coding the solution.

```python
from qiskit import *
import numpy as np
from math import pi
import matpotlib.pyplot as plt

N = 100 #Number of rotations performed
rotation_angle = pi/N #Angle of rotation
shots = 100 #Number of measurements performed

def elitzur_vaidmann_bomb(isBomb):
    """
    Input: boolean object called isBomb.
    1. isBomb = True, means the bomb is live.
    2. isBomb = False, means bomb is a dud

    Output: list of the number of measurements that
    predicted the bomb but did not detonate it, measured
    duds, and bombs that did detonate
    """


    #Performing N rotations where each measurement will
    be stored in the Classical Register
    meas = 0
```

```
#If we have a Live bomb
    #Need to keep measuring after each rotation to
reset the state back to |0>
#Bomb is a dud
    #Only need to measure once

q = QuantumRegister(1) #Setting up Quantum Register
with 1 qubit to go through the bomb
c = ClassicalRegister(meas) #Classical Register holds
 each measurement made after each rotation
circ = QuantumCircuit(q,c) #Quantum Circuit object

#Loop to perform rotations and perform measurements
after each rotation if there is a live bomb
#.....

circ.measure(0, meas-1) #For a dud, we only measure
once.

simulator = Aer.get_backend('qasm_simulator')
circ_transpile = transpile(circ, backend = simulator)
result = simulator.run(circ_transpile,shots = shots).
result()
counts = result.get_counts()

#Post-processing to make Histogram
predicted_bomb = 0
dud = 0
detonated = 0

if isBomb == True: #Bomb is alive
    predicted_bomb = counts.get('0'*meas) #predicted
bombs but no explosion
    detonated = shots - predicted_bomb - dud

else: #Bomb is a dud
    if '0' in counts:
        predicted_bomb = counts.get('0',0)
    dud = counts.get('1')
    detonated = 0

return [predicted_bomb,dud,detonated]
```

Now, use the following code to plot your results when setting the input to the function as both True and False. Plot the results as Probability (%) vs. Predicted Bomb, Dud, and Detonated:

```
predicted_bomb ,dud ,detonated = #elitzur_vaidmann_bomb
(...)
#convert results to probabilities
predicted_bomb_prob = #...
dud_prob = #...
detonated_prob = #...
plt.xticks(np.arange(3), ['Predicted Bomb', 'Dud', '
Detonated'])
plt.bar(np.arange(3),[predicted_bomb_prob ,dud_prob ,
detonated_prob])
plt.ylabel('Probability');
```

Chapter 6

Two Qubits and Entanglement

Thus far, we have seen how to manipulate, represent, and measure single qubits using linear algebra. However, a single qubit is not useful in itself for solving the difficult problems that quantum computing may be able to tackle. While superposition is an interesting property of qubits, superposition exists classically with waves. Now that we have a better understanding of how quantum mechanics works, we are ready to tackle perhaps one of the strangest phenomena in quantum physics: entanglement. This aspect of quantum computing is completely non-classical and holds the key to solving difficult problems more efficiently. The main goal of this chapter is to discuss the concept of entanglement and see how we can generate Bell states, which are the maximally entangled two-qubit states. Finally, we discuss the famous EPR paradox, which is an interesting thought experiment about entanglement, Bell's theorem and the resolution to the paradox.

6.1 Product States

We have learned that a single qubit is a two-level system, so the dimension of the complex vector space we use to work with them is $d = 2$. For single qubits, we saw that we needed two basis states to fully represent any other state in the vector space. For two qubits, the dimension of the complex vector space is $d = 4$, so we need four basis states. This is because the dimension of the complex vector space for qubits grows exponentially, as 2^n, with the number of qubits, n.

In this chapter, instead of having a 2D column vector for a single qubit, we will have a 4D column vector for two qubits.

Using the standard basis, we will have our four basis states:

$$|00\rangle = \begin{pmatrix} 1 \\ 0 \\ 0 \\ 0 \end{pmatrix} \quad |01\rangle = \begin{pmatrix} 0 \\ 1 \\ 0 \\ 0 \end{pmatrix} \quad |10\rangle = \begin{pmatrix} 0 \\ 0 \\ 1 \\ 0 \end{pmatrix} \quad |11\rangle = \begin{pmatrix} 0 \\ 0 \\ 0 \\ 1 \end{pmatrix}$$

Now, we can write down the most general two-qubit state as

$$|\psi\rangle = c_1 |00\rangle + c_2 |01\rangle + c_3 |10\rangle + c_4 |11\rangle = \begin{pmatrix} c_1 \\ c_2 \\ c_3 \\ c_4 \end{pmatrix}$$

The rules for measurement work in the same way. We need to take the squared magnitude of the projection of the basis state with the qubit state. For example, the probability to measure $|00\rangle$ is

$$\mathbb{P}(|00\rangle) = |\langle 00|\psi\rangle|^2 = \left| (1\ 0\ 0\ 0) \begin{pmatrix} c_1 \\ c_2 \\ c_3 \\ c_4 \end{pmatrix} \right|^2 = |c_1|^2$$

Similarly, we would obtain

$$\mathbb{P}(|01\rangle) = |c_2|^2 \quad \mathbb{P}(|10\rangle) = |c_3|^2 \quad \mathbb{P}(|11\rangle) = |c_4|^2$$

and again, we must satisfy the normalization condition:

$$|\,|\psi\rangle\,|^2 = |c_1|^2 + |c_2|^2 + |c_3|^2 + |c_4|^2 = 1$$

The procedure to normalize a two-qubit state is the same as before:

$$\frac{|\psi\rangle}{|\,|\psi\rangle\,|}$$

Now, let us understand how to generate these two qubits in more detail using the tensor product.

6.1.1 Tensor Product

In the two-qubit and multi-qubit case, we need to keep track of the increase in the dimension of the Hilbert space. Mathematically, we can think of joining two qubits as combining the two Hilbert spaces in some way. For each qubit, we know that each individual Hilbert space is noted as $\mathbb{C}^2$, and now, we want to go to $\mathbb{C}^4$. The operation that will take us there is called the **tensor product**.

Some information on notation and properties of the tensor product operation to keep in mind are [30]:

1. The tensor product operation is written as $\otimes$.
2. Consider two vector spaces V and W, where v is an element of V and w is an element of W. The tensor product takes the two elements v and w and generates a new element $v \otimes w$. This new element belongs to a new object $V \otimes W$, which is called a **tensor**. The dimension of the tensor is the product of the dimensions of the two vector spaces V and W.
3. Recall our discussion about basis states from previous chapters, let's suppose that the set of vectors $\{v_i\}$ are a set of basis vectors for V and the set of vectors $\{w_i\}$ are a set of basis vectors for W, the basis for $V \otimes W$ is actually the set $\{v_i \otimes w_i\}$.
4. The tensor product operation is associative implying that

$$(u \otimes v) \otimes w = u \otimes (v \otimes w)$$

5. The tensor product is not commutative implying that

$$u \otimes v \neq v \otimes u$$

Computing the Tensor Product

Having some familiarity with the properties of the tensor product, let us use it to derive a general two-qubit state. We know that each qubit belongs to a 2D complex vector space denoted as $\mathbb{C}^2$, thus two qubits belong to $\mathbb{C}^2 \otimes \mathbb{C}^2$, which is a tensor with dimension four.

Let's suppose we have the following two qubits:

$$|u\rangle = \begin{pmatrix} a \\ b \end{pmatrix} \qquad |v\rangle = \begin{pmatrix} c \\ d \end{pmatrix}$$

To derive the **product state**, $|uv\rangle$, we perform the tensor product as follows:

$$|uv\rangle = |u\rangle \otimes |v\rangle = \begin{pmatrix} a \times \begin{pmatrix} c \\ d \end{pmatrix} \\ b \times \begin{pmatrix} c \\ d \end{pmatrix} \end{pmatrix} = \begin{pmatrix} ac \\ ad \\ bc \\ bd \end{pmatrix}$$

The tensor product is not a commutative operation, so the state $|vu\rangle$ is different:

$$|vu\rangle = |v\rangle \otimes |u\rangle = \begin{pmatrix} c \times \begin{pmatrix} a \\ b \end{pmatrix} \\ d \times \begin{pmatrix} a \\ b \end{pmatrix} \end{pmatrix} = \begin{pmatrix} ca \\ cb \\ da \\ db \end{pmatrix}$$

We can also perform a tensor product of matrices. Let's take some matrices we are familiar with from Chapter 4. For example, suppose we wanted to take the tensor product of a Pauli X-gate with a Pauli Z-gate:

$$X \otimes Z = \begin{pmatrix} 0 & 1 \\ 1 & 0 \end{pmatrix} \otimes \begin{pmatrix} 1 & 0 \\ 0 & -1 \end{pmatrix} = \begin{pmatrix} 0 \times \begin{pmatrix} 1 & 0 \\ 0 & -1 \end{pmatrix} & 1 \times \begin{pmatrix} 1 & 0 \\ 0 & -1 \end{pmatrix} \\ 1 \times \begin{pmatrix} 1 & 0 \\ 0 & -1 \end{pmatrix} & 0 \times \begin{pmatrix} 1 & 0 \\ 0 & -1 \end{pmatrix} \end{pmatrix}$$

$$X \otimes Z = \begin{pmatrix} 0 & 0 & 1 & 0 \\ 0 & 0 & 0 & -1 \\ 1 & 0 & 0 & 0 \\ 0 & -1 & 0 & 0 \end{pmatrix}$$

Now, let's use this new matrix we have constructed and operate on the $|01\rangle$ state that we introduced in the previous section. Let us call $X \otimes Z$, XZ, which is the lazy notation so that we don't always need to write the tensor product. We know that $|01\rangle = |0\rangle \otimes |1\rangle = \begin{pmatrix} 0 \\ 1 \\ 0 \\ 0 \end{pmatrix}$, so $XZ\,|01\rangle$ is

$$XZ\,|01\rangle = X{\otimes}Z(|0\rangle{\otimes}|1\rangle) = \begin{pmatrix} 0 & 0 & 1 & 0 \\ 0 & 0 & 0 & -1 \\ 1 & 0 & 0 & 0 \\ 0 & -1 & 0 & 0 \end{pmatrix}\begin{pmatrix} 0 \\ 1 \\ 0 \\ 0 \end{pmatrix} = \begin{pmatrix} 0 \\ 0 \\ 0 \\ -1 \end{pmatrix} = -\,|11\rangle$$

It appears as though the X-gate acted on the first qubit and the Z-gate acted on the second qubit. In mathematical terms, we want to verify the following:

$$X \otimes Z(|0\rangle \otimes |1\rangle) \overset{?}{=} (X\,|0\rangle) \otimes (Z\,|1\rangle)$$

From the right-hand side of the above equation, we know that

$$X\,|0\rangle = \begin{pmatrix} 0 & 1 \\ 1 & 0 \end{pmatrix}\begin{pmatrix} 1 \\ 0 \end{pmatrix} = \begin{pmatrix} 0 \\ 1 \end{pmatrix} = |1\rangle$$

and

$$Z\,|1\rangle = \begin{pmatrix} 1 & 0 \\ 0 & -1 \end{pmatrix}\begin{pmatrix} 0 \\ 1 \end{pmatrix} = \begin{pmatrix} 0 \\ -1 \end{pmatrix} = -\,|1\rangle$$

So,

$$(X\,|0\rangle) \otimes (Z\,|1\rangle) = |1\rangle \otimes (-\,|1\rangle) = -\,|1\rangle \otimes |1\rangle = -\,|11\rangle$$

Thus, we have shown that

$$X \otimes Z(|0\rangle \otimes |1\rangle) = (X\,|0\rangle) \otimes (Z\,|1\rangle)$$

In general, if we have some gates A and B, and two quantum states $|u\rangle$ and $|v\rangle$,

$$(A \otimes B)(|u\rangle \otimes |v\rangle) = (A\,|u\rangle) \otimes (B\,|v\rangle)$$

This is a useful mathematical property of tensor products which saves some time in computing multiplication of vectors and matrices of higher dimensions. In the case of two-qubit product states, we can compute the result of each gate on the specific qubit it operates on and this is determined by the order in which the gates are written. As we saw in the above example, XZ means that X acts on the first qubit and Z acts on the second qubit. We can also have two-qubit operations that only change the state of one qubit. For example, IX means identity is applied to the first qubit and X is applied to the second qubit. As we saw before, the identity operation just gives you back the same state, so the IX-gate is a two-qubit gate that only changes the second qubit.

Checkpoint Exercises

1. Show using the tensor product operation that the vectors $|00\rangle$, $|01\rangle$, $|10\rangle$, and $|11\rangle$ are

$$|00\rangle = \begin{pmatrix} 1 \\ 0 \\ 0 \\ 0 \end{pmatrix} \quad |01\rangle = \begin{pmatrix} 0 \\ 1 \\ 0 \\ 0 \end{pmatrix} \quad |10\rangle = \begin{pmatrix} 0 \\ 0 \\ 1 \\ 0 \end{pmatrix} \quad |11\rangle = \begin{pmatrix} 0 \\ 0 \\ 0 \\ 1 \end{pmatrix}$$

2. Check if the following states are normalized. If they are not, normalize them:

$$|\psi\rangle = \tfrac{1}{2}(|00\rangle + |01\rangle + |10\rangle + |11\rangle)$$
$$|\psi\rangle = \tfrac{1}{2}|00\rangle + \tfrac{1}{4}|01\rangle$$

3. Perform the following tensor products:

$$X \otimes Z(|0\rangle \otimes |+\rangle)$$
$$I \otimes Y(|1\rangle \otimes |-\rangle)$$

6.2   Entangled States

The product we saw in the previous section are not entangled because we are able to perform independent measurements on the qubit, where knowledge from measuring one qubit does not tell you anything about the others. In other words, measurements are not correlated.

However, quantum states also have an interesting property in that they may become entangled. **Quantum entanglement** occurs when a group of particles are either generated from source, interacted in some way, or are near each other spatially so that each particle in the group is not independent of the state of the others. This means that measurements are correlated. In fact, entangled particles may be separated from each other by a large distance and the measurement would still be correlated. Entanglement is a phenomenon that not only highlights the major differences between classical and quantum physics but also brings up many philosophical questions about locality.[1]

6.2.1 Physical Generation of Entangled States

Physically generating entangled states still remains an open area of research. Researchers are curious about new and useful ways to generate entanglement for potentially realizing quantum networks. Quantum networks rely primarily on efficiently generating entanglement between qubits and photons since photons can travel for long distances.

One way to generate entangled photons is using a process known as spontaneous parametric down-conversion or SPDC [31]. The process typically relies on a nonlinear crystal which splits a single photon typically called a *pump* photon into pairs of photons called *signal* and *idler* photons. The "spontaneous" part of the name is because there is no applied field that stimulates the process. "Parametric" refers to a process that depends on the electric field which is a vector quantity. "Down-conversion" refers to the fact that a photon splits into two lower energy photons. So, in simpler terms, this process takes in a single photon and produces two photons that must satisfy energy and momentum conservation.

The conservation of energy and momentum are related to the frequency, ω, and propagation vectors, $\vec{k}$, of the input photon and output photons by Planck's constant, respectively:

$$\hbar\omega_p = \hbar\omega_1 + \hbar\omega_2$$

$$\hbar\vec{k}_p = \hbar\vec{k}_1 + \hbar\vec{k}_2$$

These equations are known as the *phase matching conditions* and they tell us the frequency of the incoming light to the crystal must be equal to sum of the frequencies of the two outgoing photons and similarly for $\vec{k}$, the vector of propagation. The frequencies and directions of propagation for the outgoing photons depends on the properties of the nonlinear crystal, such as the index of refraction and angle

---

[1]In simple terms, the principle of locality means that an object is influenced by its immediate environment, which entanglement seems to violate.

of incidence of the incoming photon. Inside this crystal, the light is
reflected and refracted in complicated ways, which we will not go
into details here. The picture to keep in mind is that by sending in
a pump photon into the nonlinear crystal, we get an entangled pair
of photons (signal and idler).

When both of the outgoing photons have the same polarization,
this is called Type I down-conversion.[2] Recalling from Section 3.1.2
on light polarization, $|H\rangle$ corresponds to a horizontally polarized
photon, and $|V\rangle$ refers a vertically polarized photon, the general state
will be

$$|\Phi\rangle = \frac{1}{\sqrt{2}}\left(|H_1 H_2\rangle + e^{i\varphi}|V_1 V_2\rangle\right)$$

When the outgoing photons do not have the same polarization,
this is called Type II down-conversion, then the general state will be

$$|\Psi\rangle = \frac{1}{\sqrt{2}}\left(|H_1 V_2\rangle + e^{i\varphi}|V_1 H_2\rangle\right)$$

These states are highly correlated and cannot be expressed as a product state. This fact is what leads to the interpretation that an entangled state cannot be fully described by the state of one of the photons;
the states of both photons are needed. In this case, it is the nonlinear crystal that causes the signal and the idler to become entangled.
Experimentally, the special orientations of the crystal's optical axis
are studied so that the incoming photon source (pump photons) can
be adjusted to generate outgoing photons that are entangled. To
deduce whether they are entangled, the correlations of the two outgoing photons are measured. This experiment would be done many
times and statistics would be collected in order to see this. In the
Type I down-conversion case, if one of the photons is either ordinary

---

[2]Type I spontaneous down-conversion occurs when the signal and the idler both
have ordinary polarization. In order to get correlated states, two nonlinear crystals
are used and they are oriented so that from the perspective of the pump photon, one of the crystals is rotated 90° about the direction of propagation of the
pump photon. Essentially, this is creating the horizontal (ordinary) and vertical
(extraordinary) basis, so now the pump appears as though it's in a superposition
of ordinary and extraordinary polarizations when it interacts with the crystal.

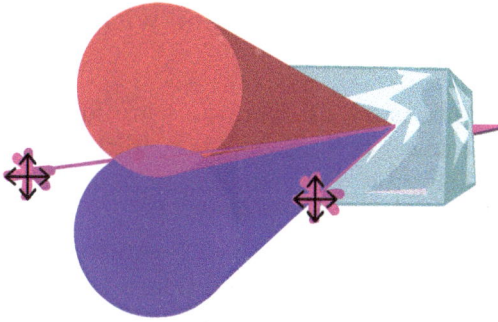

Figure 6.1. Entangled photons generated by sending coherent laser light through a nonlinear crystal.

or extraordinary, the other will also be the same. In the Type II down-conversion, if one of the photons is ordinary, the other will be extraordinary and vice versa. Through many measurements, the statistics tells experimenters if they are really generating entangled photons or not.

For example, the result for Type II down-conversion can be understood from Fig. 6.1. The generated photons are emitted in two cones. The top and bottom cones have two different polarizations and the additional phase φ is due to birefringence from the nonlinear crystal which introduces a relative phase between the ordinary (typically considered to be horizontal polarization) and extraordinary polarized light (typically considered to be vertical polarization). In this case, photons are found in a superposition where the signal and idler cones intersect. At these intersection points, it is indeterminate what the polarization of a photon was before a measurement was made and this is why a superposition state results from the above equations.

Checkpoint Exercise

In Quantum Flytrap (see Appendix B), it is possible to create entangled states using a BBO crystal which is an acronym for Beta Barium Borate, a type of nonlinear optical crystal. Right click on the crystal, select the different types of Bell states (labeled $|\Psi_\pm\rangle$ and $|\Phi_\pm\rangle$) that can be generated, and using the "Waves" view and the ket representation on the right-hand side of window, write down the Bell states in the photon polarization basis.

6.2.2   Generating Entanglement with Controlled Quantum Gates

Controlled quantum gates or operations are formed by applying a gate to one qubit where the outcome is conditioned or dependent on the state of another qubit. For example, let's say you want to flip the first qubit *if* the second qubit is in the $|1\rangle$ state. This operation is known as a controlled X-gate or CNOT gate because it performs a Pauli X-gate on the second qubit conditioned on the first qubit being in the $|1\rangle$ state. In fact, this gate corresponds to the XOR logic gate in classical computing. This is just one example of a controlled gate. In general, you can have a controlled gate that applies some arbitrary rotation to the first qubit outcome conditioned on the second qubit.

Let us study the CNOT gate by using the truth table of the gate, which tells us the outputs for some given inputs. Suppose that we have two qubits, we saw before that the complete basis of two qubits consists of the four basis states $\{|00\rangle, |01\rangle, |10\rangle, |11\rangle\}$. If we want to derive a matrix of the CNOT gate, we have to consider what this gate will do to each of these basis states. Consider that this gate flips the second qubit only if the first qubit is in the $|1\rangle$ state.

As we can see from the truth table in Table 6.1, we may now express the CNOT gate as a matrix in the two qubit basis $\{|00\rangle, |01\rangle, |10\rangle, |11\rangle\}$:

$$\text{CNOT} = \begin{pmatrix} 1 & 0 & 0 & 0 \\ 0 & 1 & 0 & 0 \\ 0 & 0 & 0 & 1 \\ 0 & 0 & 1 & 0 \end{pmatrix}$$

Table 6.1.   Truth table for the CNOT gate.

Before CNOT	After CNOT
$|00\rangle$	$|00\rangle$
$|01\rangle$	$|01\rangle$
$|10\rangle$	$|11\rangle$
$|11\rangle$	$|10\rangle$

In these control gates, the qubit that we do an operation on (in the example above, the second qubit) is called the **target** qubit. The qubit that controls whether we do an operation to the target qubit is conveniently called the **control** qubit.

We can use the same procedure to derive other controlled Pauli operations or controlled rotations about the x-, y-, or z-axes or any arbitrary axis. So, if we keep the control qubit to be the first qubit and the target to be the second qubit, then we can generate any controlled operation in a similar way:

$$|00\rangle \mapsto |00\rangle$$
$$|01\rangle \mapsto |01\rangle$$
$$|10\rangle \mapsto |1\rangle \otimes U|0\rangle$$
$$|11\rangle \mapsto |1\rangle \otimes U|1\rangle$$

So, if we have some arbitrary single-qubit unitary gate parameterized in terms of angles, $U(\theta, \phi, \lambda)$, then we would obtain the following controlled unitary gate, CU:

$$CU(\theta, \phi, \lambda) = \begin{pmatrix} 1 & 0 & 0 & 0 \\ 0 & 1 & 0 & 0 \\ 0 & 0 & \cos(\theta/2) & e^{-i\lambda}\sin(\theta/2) \\ 0 & 0 & e^{i\varphi}\sin(\theta/2) & e^{i(\lambda+\varphi)}\cos(\theta/2) \end{pmatrix}$$

We can now see the structure of the controlled operation. If the control qubit is $|0\rangle$, we apply the identity gate as can be seen in the first two rows and columns of the matrix. If the control qubit is $|1\rangle$, we apply the gate of interest to the target qubit.

6.2.3 Bell States

Now that we've learned more about how entangled states could be physically generated and have these new controlled gates in our arsenal, let us examine how to generate maximally entangled states.

We need to make use of the controlled gates that we have learned. Recall in our discussion of spontaneous parametric down-conversion that the way to generate entangled states is by having the pump photon in a superposition state when it enters the nonlinear crystal. The superposition state and crystal orientation would then determine which one of the maximally entangled states would be measured. Now, in a more mathematical picture, the nonlinear crystal is really serving as some sort of controlled operation.

Let the ordinary polarization be denoted by $|0\rangle$ and the extraordinary polarization by $|1\rangle$. Now, let's focus on one of the controlled operations we just learned, say CNOT. We know that for this operation, we need a control and target qubit, where the control basically signals whether or not to do an operation on the target. We will continue to stick with the convention that if the control qubit is $|1\rangle$, we perform the operation on the target qubit.

Let's try the case when the control qubit is in an equal superposition state and the target is in state $|0\rangle$:

$$|\psi_{\text{Control}}\rangle = \frac{1}{\sqrt{2}}\left(|0\rangle + |1\rangle\right)$$

$$|\psi_{\text{Target}}\rangle = |0\rangle$$

$$\left|\tilde{\psi}\right\rangle = |\psi_{\text{Control}}\rangle \otimes |\psi_{\text{Target}}\rangle = \frac{1}{\sqrt{2}}\left(|00\rangle + |10\rangle\right) = \frac{1}{\sqrt{2}}\begin{pmatrix} 1 \\ 0 \\ 1 \\ 0 \end{pmatrix}$$

Let's take this state and apply CNOT:

$$\text{CNOT}\left|\tilde{\psi}\right\rangle = \begin{pmatrix} 1 & 0 & 0 & 0 \\ 0 & 1 & 0 & 0 \\ 0 & 0 & 0 & 1 \\ 0 & 0 & 1 & 0 \end{pmatrix} \frac{1}{\sqrt{2}}\begin{pmatrix} 1 \\ 0 \\ 1 \\ 0 \end{pmatrix} = \frac{1}{\sqrt{2}}\begin{pmatrix} 1 \\ 0 \\ 0 \\ 1 \end{pmatrix}$$

So, we have ended up with the following state:

$$|\Phi_+\rangle = \frac{1}{\sqrt{2}}\left(|00\rangle + |11\rangle\right)$$

This is an example of a Bell state. Looking back at the spontaneous down-conversion section, we can compare and see that this is one state that results from Type I down-conversion. This is a maximally entangled state because there is no way that we can decompose this state into a superposition of product states. We can also generate the antisymmetric state by having the control as

$$|\psi_{\text{Control}}\rangle = \frac{1}{\sqrt{2}} (|0\rangle - |1\rangle)$$

so that we obtain the following state when applying CNOT:

$$|\Phi_-\rangle = \frac{1}{\sqrt{2}} (|00\rangle - |11\rangle)$$

The other two Bell states are

$$|\Psi_\pm\rangle = \frac{1}{\sqrt{2}} (|01\rangle \pm |10\rangle)$$

You will figure out how to generate these in the homework exercises at the end of the chapter.

6.2.4 Quantifying the Degree of Entanglement

We know about how to generate maximally entangled states, but what about a state that might not be maximally entangled, how can we quantify its degree of entanglement? Well, one simple measure is called **concurrence**. Suppose we have a general two-qubit state:

$$|\psi\rangle = c_1 |00\rangle + c_2 |01\rangle + c_3 |10\rangle + c_4 |11\rangle$$

The concurrence is defined as

$$C = 2|c_1 c_4 - c_2 c_3|$$

Note that for a maximally entangled state, like Bell states, the concurrence will reach the maximum value of $C = 1$.

Now, suppose we have some state:

$$|\psi\rangle = \frac{1}{\sqrt{3}} (|00\rangle + |01\rangle + |11\rangle)$$

Then its concurrence would be

$$C = 2 \left| \frac{1}{\sqrt{3}} \cdot \frac{1}{\sqrt{3}} - \frac{1}{\sqrt{3}} \cdot 0 \right| = \frac{2}{3}$$

We can see here that this state, $|\psi\rangle$, almost looks like a Bell state, but the correlation is not perfect because we have some component of $|01\rangle$. So, this state is only partially entangled.

Meanwhile, for a product state, we would have all the basis states in the superposition, for example,

$$|+\rangle \otimes |-\rangle = \frac{1}{2}\left(|00\rangle - |01\rangle + |10\rangle - |11\rangle\right) \qquad (6.1)$$

Here, the concurrence would be

$$C = 2\left| -\frac{1}{2} \cdot \frac{1}{2} - \left(-\frac{1}{2} \cdot \frac{1}{2}\right)\right| = 0$$

Checkpoint Exercise

1. Using the Qiskit Quantum Circuits class, generate and visualize all of the Bell states we discussed in the previous section. The CNOT gate in Qiskit can be implemented as follows:

```
q = QuantumRegister (2)
c = Classical Register (2)
circ.cx(q[0], q[1]) #CNOT gate, where the first
argument specifies the control qubit and the second
argument specifies the target qubit
```

As we are working with two qubits, the Bloch sphere will not be useful anymore since it is only a single-qubit representation. There's no nice way to visualize 4D complex space, so we will instead visualize Bell states using the graphing tool:

```
plot_state_city
```

This plot is introducing a topic that we will not cover in this introductory book, which is density matrix formalism. It is yet another way to represent quantum states through a matrix instead of simply the state vector. This representation allows us to write down states that are not pure, implying that they cannot be expressed in terms of a state vector.

2. Compute the concurrence for the following states:

   (a) $|\psi\rangle = \frac{1}{2}\left(\sqrt{2}\,|00\rangle - |01\rangle + |10\rangle\right)$,

   (b) $|\psi\rangle = \frac{1}{\sqrt{3}}\left(\sqrt{2}\,|00\rangle + |11\rangle\right)$,

   (c) $|\psi\rangle = \frac{1}{\sqrt{2}}\left(|01\rangle + e^{i\phi}\,|10\rangle\right)$.

6.3 EPR Paradox

The early 20th century was filled in exciting theoretical and experimental discoveries which contributed to the development of quantum theory. Planck's work on blackbody radiation, Einstein's Nobel Prize winning work on the photoelectric effect and Bohr's revolutionary paper on the hydrogen atom spurred the important development of the probabilistic interpretation of quantum mechanics. This involved Schrödinger's equation which was developed to describe the evolution of the wavefunction. Further, Heisenberg formulated the uncertainty principle and matrix mechanics to highlight that operators and not the wavefunction itself evolves with time. The uncertainty principle states that observables that do not commute[3] cannot have simultaneous eigenstates, implying that you cannot have the "simultaneous realities" from the resulting measurement of the two observables. For example, if a quantum state is localized in space, implying that it was measured and found to be in some particular location, you have maximal uncertainty about its speed or momentum because position and momentum are non-commuting observables. Einstein was a firm opponent of the uncertainty principle and opposed even more the idea of the non-locality that was permitted by considering entangled states. As we learned, non-locality is at the heart of entanglement since the states of two entangled particles are dependent on each other regardless of how much they are spatially separated.

The EPR paradox is based on a paper published in *Physical Review* by Einstein, Podolsky, and Rosen in 1935 at Princeton University [13]. This paper brought to question the completeness of quantum mechanics, specifically, it questions the key claim of quantum theory: the wavefunction of the system provides a complete description of the system. Their paper defines "physical reality" to consist of physical quantities which can be predicted *with certainty* without needing to disturb the system. The suggested solution to the conundrum is whether there are some *hidden variables* that determine state collapse.

---

[3]The commutator of two observables in this case tells us whether multiplication of two observables **A** and **B** is commutative, i.e. is **AB = BA**?

The setup of the thought experiment goes something like this:

> Suppose that we have a neutral π-meson in our large particle detector and we observe that it quickly (in about 850 fs) decays into a pair of photons. This pair of photons can be described by a single quantum state, so the photons are entangled with each other. As time passes, the photons continue to move (in opposite directions with opposite spins because of linear and angular momentum conservation), but they *remain* entangled nonetheless. Let's label one photon A and the other B and suppose the experimenter measures the spin in z-direction of the photon A. If we measure A to be spin-up, then they know that B must have spin-down without ever needing to measure B. Now, we learned before that measuring in z will project the spin along the z-axis and the spin will be either up or down. Since A has been projected into $+z$, if we try to measure spin-x, we have a 50–50 chance to get spin-up or spin-down along x-axis and conditioned on that the spin of B will be the opposite. It seems as though the measurement of A is *instantaneously* telling us about the state of B.

The seemingly *instantaneous* nature of the measurement of A telling us what B is really disturbed Einstein about entanglement. He questioned the validity of entanglement because it seemed to contradict a key postulate of the special theory of relativity, namely that speed of light is constant in all reference frames. This deep distrust came from not accepting non-locality. Essentially, if entanglement was local, this means that something would have to travel through space instantly and tell B what the spin of A was measured to be so that B can be the opposite of that, which is not physically plausible and contradicts special relativity! This is also where the phrase "spooky action at a distance" was coined. Due to how the problem was outlined, there were basically two options on the table:

1. Accept quantum mechanics despite the discomforting contradiction with special relativity.
2. Propose that there are *local* hidden variables that solve the problem of seemingly instantaneous action at a distance and satisfy special relativity.

You can guess which one Einstein went with!

Bell's Theorem

For a while, this thought experiment sat with scientists, unproven, and the problem was rephrased in different ways which all seemed to point to how quantum states cannot somehow represent physical reality. Meanwhile, quantum mechanics was proving itself to be useful and yielding correct predictions for experimental results. It formed the basis for explaining quantum chemistry, electronics, and quantum optics. Of particular importance to technology, having a good understanding of semiconductor physics led to the invention of the transistor. So, while this seemingly uncomfortable point about implications of non-locality was still in the air, quantum mechanics continued to be useful in understanding our world.

In 1964, John Bell tried to clear the air [14]. He proposed a way to test for the existence of hidden variables and developed an inequality or bound that could tell experimenters about hidden variables based on the statistics of their experiment. In fact, Bell initially studied whether it was possible to solve the non-locality "issue" with quantum mechanics using hidden-variable theories and in fact he found that you could, but only for a specific case of the problem. The general problem could not be explained with hidden-variable theories. In fact, he concluded that *any* local hidden variable theory cannot explain the predictions of quantum mechanics.

To understand the first part of Bell's theorem, let's start by assuming we have the following entangled state:

$$|\psi\rangle = \frac{|\uparrow\downarrow\rangle - |\downarrow\uparrow\rangle}{\sqrt{2}}$$

This is known as a spin singlet. We can see here that the two particles are anti-correlated so that when particle 1 is spin-up, particle 2 is spin-down, and vice versa. Instead of thinking about the case of measuring the spin of both particles along the same direction (e.g. $\hat{z}$ as we considered in the EPR paradox), Bell thought about a more general case where we could measure the spin of particle A along some arbitrary axis $\vec{a}$ and the spin of particle B along another axis $\vec{b}$. We can then gather statistics from such an experiment by recording +1 for spin-up and −1 for spin-down along the axes we are measuring

Table 6.2. Statistics from measuring the spins of Particle A and B along axes $\vec{a}$ and $\vec{b}$, respectively.

Particle A	Particle B
+1	−1
−1	−1
+1	−1
−1	+1
+1	+1
⋮	⋮

for each particle. What we are doing here is building a probability distribution for measuring spin along the specified axis. Our statistics will then look something like that in Table 6.2.

To have a compact way of representing our results, we can think about the measurement along each axis as a random variable, $\vec{\sigma_a}$ and $\vec{\sigma_b}$, denoting spin along $\vec{a}$ and $\vec{b}$ axes, respectively. If we assume that these random variables are *independent*, then the average value of $\vec{\sigma_a} \cdot \vec{\sigma_b}$, or expectation value, can be denoted as

$$E(\vec{a}, \vec{b}) = \langle \vec{\sigma_a} \cdot \vec{\sigma_b} \rangle = -\vec{a} \cdot \vec{b}$$

We get back to the EPR paradox problem when we assume that $\vec{a} = \vec{b}$, and thus, $E(\vec{a}, \vec{b}) = -1$ always for the spin singlet state because if we measure spin-up for particle A, then we will always measure spin-down for particle B and vice versa.

The following is a derivation of Bell's famous inequality. Let us suppose that the outcomes of measuring particles A and B along their respective axes depend on a hidden variable λ which we do not control. We assume that it has some probability distribution associated with it $\Lambda(\lambda)$ which is normalized so that $\sum_\lambda \Lambda(\lambda) = 1$. The measurement of particle A is now described as some function of $\vec{a}$ and λ, $A(\vec{a}, \lambda)$ and similarly for particle B, $B(\vec{b}, \lambda)$. Now, our average value is[4]

$$E(\vec{a}, \vec{b}) = \sum_\lambda \Lambda(\lambda) A(\vec{a}, \lambda) B(\vec{b}, \lambda)$$

---

[4]Recall our discussion in Section 1.2.4 about calculating expected values.

It is still always true that if we have the spin singlet state and $\vec{a} = \vec{b}$, then the results are anti-correlated so that $A(\vec{a}, \lambda) = -B(\vec{a}, \lambda)$ or $A(\vec{b}, \lambda) = -B(\vec{b}, \lambda)$

Thus, we can write the expected value as

$$E(\vec{a}, \vec{b}) = -\sum_{\lambda} \Lambda(\lambda) A(\vec{a}, \lambda) A(\vec{b}, \lambda)$$

Now, suppose that we measure the spins of the particles along another axis $\vec{c}$ for which the measurement outcome will now be $A(\vec{c}, \lambda) = -B(\vec{c}, \lambda)$, then

$$E(\vec{a}, \vec{b}) - E(\vec{a}, \vec{c}) = -\sum_{\lambda} \Lambda(\lambda) \left(A(\vec{a}, \lambda) A(\vec{b}, \lambda) - A(\vec{a}, \lambda) A(\vec{c}, \lambda) \right)$$

Factoring out $A(\vec{a}, \lambda) A(\vec{b}, \lambda)$ and using the fact that $(A(\vec{b}, \lambda))^2 = 1$, we obtain

$$E(\vec{a}, \vec{b}) - E(\vec{a}, \vec{c}) = -\sum_{\lambda} \Lambda(\lambda) \left(1 - A(\vec{b}, \lambda) A(\vec{c}, \lambda) \right) A(\vec{a}, \lambda) A(\vec{b}, \lambda)$$

We also know that $|A(\vec{a}, \lambda) A(\vec{b}, \lambda)| = 1$, so let's take the absolute value of both sides of the equation and also use the fact that $\Lambda(\lambda) \left(1 - A(\vec{b}, \lambda) A(\vec{c}, \lambda) \right) \geq 0$, so we get that

$$|E(\vec{a}, \vec{b}) - E(\vec{a}, \vec{c})| \leq \sum_{\lambda} \Lambda(\lambda) \left(1 - A(\vec{b}, \lambda) A(\vec{c}, \lambda) \right)$$

We know that $\sum_{\lambda} \Lambda(\lambda) = 1$ and $\sum_{\lambda} \Lambda(\lambda) A(\vec{b}, \lambda) A(\vec{c}, \lambda) = E(\vec{b}, \vec{c})$

So, we finally have that

$$|E(\vec{a}, \vec{b}) - E(\vec{a}, \vec{c})| \leq 1 + E(\vec{b}, \vec{c})$$

As you can see, the derivation of Bell's theorem is relatively simple and the result is powerful. What we will see is that predictions from quantum mechanics violate this inequality. Further, experiments testing the inequality are consistent with quantum mechanics and also violate the inequality. Physicists Alain Aspect, John F. Clauser, and Anton Zeilinger won the 2022 Nobel Prize in Physics for experimentally proving that Bell's inequality is violated.

Let's do a concrete example to understand how it works by considering the Stern–Gerlach experiment for measuring electron spin back in Section 3.1.1. By passing a beam of electrons through a detector aligned along z, for example, we will observe the particles deflect up or down. So, you can measure the spin to be either $+\frac{\hbar}{2}$ or $-\frac{\hbar}{2}$ along any axis. We have also seen through examples in that section that it's not possible to measure spin in z if your state is an eigenstate of spin in x and vice versa. If the state is in some definite spin along one axis, it's not possible to measure it along another axis. For a detector aligned along x, we would observe left and right. In the electron spin basis, the four Bell states would be

$$|\Phi_{\pm}\rangle = \frac{1}{\sqrt{2}} \left(|\uparrow\uparrow\rangle \pm |\downarrow\downarrow\rangle \right)$$

$$|\Psi_{\pm}\rangle = \frac{1}{\sqrt{2}} \left(|\uparrow\downarrow\rangle \pm |\downarrow\uparrow\rangle \right)$$

The original formulation considers the following entangled state, in which the spins of the electrons are anti-correlated:

$$|\Psi_{-}\rangle = \frac{|\uparrow\downarrow\rangle - |\downarrow\uparrow\rangle}{\sqrt{2}}$$

Suppose that (somehow) we are able to take these entangled electrons and separate them so that one is in a lab in the US and the other in Europe. In both labs, they have Stern–Gerlach detectors that can be rotated to arbitrary angles, and the three angles that are agreed upon are angles A, B, and C to be $0°, 45°$ and $90°$. When measuring along each axis, the outcome will be that the electron is spin-up or spin-down. The possible states that each electron can be in after measurement is summarized in the Venn diagram in Fig. 6.2. We can see that for any pair of measurements of the electrons, two labs always obtain anti-correlated results regardless of the axis that the spin is measured along.

At least $\frac{2}{3}$ of the time, the labs will obtain the same results in any one experiment where they measure along different axes. For example, if one of the electrons is measured along $0°$ and the other is $45°$ in one lab and in the other $45°$ and $0°$, then we can see that $\frac{2}{3} \approx 67\%$ of the time they measure $\rightarrow$ for in the $0°$ detector and $\nearrow$ in the $45°$. However, if we look at all the statistics of this case, for

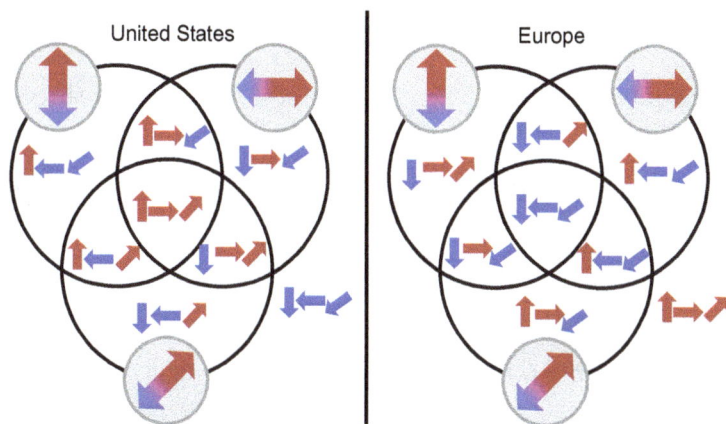

Figure 6.2. Two Venn diagrams representing the eight possible states that the electrons can have after going through the three different detector orientations in the United States lab and the Europe lab. The three circles represent the orientation of the electron and the elements inside the circle are the possible measurement outcomes for each detector orientation. Red represents the up, right, or diagonal eigenstate, and blue represents the down, left, and anti-diagonal eigenstate depending on the basis.

the different spin orientations, we find that $\frac{6}{8}$ or 75% of the time we get the same result if the electrons are measured along different axes in the two labs, which is greater than Bell's upper bound!

Checkpoint Exercise

1. Go through the steps we took to derive Bell's theorem once more and show that from

$$E(\vec{a},\vec{b}) - E(\vec{a},\vec{c}) = -\sum_\lambda \Lambda(\lambda)\left(A(\vec{a},\lambda)A(\vec{b},\lambda) - A(\vec{a},\lambda)A(\vec{c},\lambda)\right)$$

we end up with

$$E(\vec{a},\vec{b}) - E(\vec{a},\vec{c}) = -\sum_\lambda \Lambda(\lambda)\left(1 - A(\vec{b},\lambda)A(\vec{c},\lambda)\right)A(\vec{a},\lambda)A(\vec{b},\lambda)$$

Note that $A(\vec{a},\lambda) = \pm 1$, $A(\vec{b},\lambda) = \pm 1$ and $A(\vec{c},\lambda) = \pm 1$ so $(A(\vec{a},\lambda))^2 = 1$ and so on.

6.4   Three-Qubit States and Operations

For three qubits, we have $2^3 = 8$ basis states. The general three-qubit state can be written as

$$|\psi\rangle = c_1 |000\rangle + c_2 |001\rangle + c_3 |010\rangle + c_4 |011\rangle + c_5 |100\rangle + c_6 |101\rangle$$

$$+ c_7 |110\rangle \, c_8 |111\rangle = \begin{pmatrix} c_1 \\ c_2 \\ c_3 \\ c_4 \\ c_5 \\ c_6 \\ c_7 \\ c_8 \end{pmatrix}$$

where $|000\rangle = |0\rangle \otimes |0\rangle \otimes |0\rangle$, $|001\rangle = |0\rangle \otimes |0\rangle \otimes |1\rangle$, and so on.

Three-qubit operations can be constructed in a similar way by taking tensor products of single-qubit gates, e.g. $IZX = I \otimes Z \otimes X$, where

$$IZX |101\rangle = (I |1\rangle) \otimes (Z |0\rangle) \otimes (X |1\rangle) = |100\rangle$$

There is also an important three-qubit gate known as a Toffoli gate which you can also call a controlled CNOT gate or CCNOT. This gate will perform an X-gate on the third qubit, only if the first two qubits are in $|1\rangle$. So, if we were to write down the matrix in the three-qubit basis we just discussed (keeping the ordering as listed in the general qubit state), we would get

$$\text{TOFF} = \text{CCNOT} = \begin{pmatrix} 1 & 0 & 0 & 0 & 0 & 0 & 0 & 0 \\ 0 & 1 & 0 & 0 & 0 & 0 & 0 & 0 \\ 0 & 0 & 1 & 0 & 0 & 0 & 0 & 0 \\ 0 & 0 & 0 & 1 & 0 & 0 & 0 & 0 \\ 0 & 0 & 0 & 0 & 1 & 0 & 0 & 0 \\ 0 & 0 & 0 & 0 & 0 & 1 & 0 & 0 \\ 0 & 0 & 0 & 0 & 0 & 0 & 0 & 1 \\ 0 & 0 & 0 & 0 & 0 & 0 & 1 & 0 \end{pmatrix}$$

An important three-qubit state is the GHZ (Greenberger–Horne–Zelinger) state. The GHZ state is basically a maximally entangled three qubit state given as follows:

$$|\text{GHZ}\rangle = \frac{|000\rangle + |111\rangle}{\sqrt{2}}$$

To generate the state from an initial state $|000\rangle$, first we perform a Hadamard gate on the first qubit and then a string of CNOTs between the first and second qubits and the second and third qubits. You may go through the exercise of deriving the state. This state is particularly interesting because if we were to measure the third particle of the GHZ state, we could end up with a maximally entangled Bell state. You can also prove that $|\text{GHZ}\rangle$ state can be written as

$$|\text{GHZ}\rangle = \frac{1}{2}\left(|00\rangle + |11\rangle\right) \otimes |+\rangle + \frac{1}{2}\left(|00\rangle - |11\rangle\right) \otimes |-\rangle$$

So, you can see that if we were to measure the third particle in the $\{|+\rangle, |-\rangle\}$ basis, we would end up with one of the Bell states. For more than three qubits, we can still construct GHZ-like states as follows:

$$|\text{GHZ}\rangle = \frac{|000\ldots0\rangle + |111\ldots1\rangle}{\sqrt{2}}$$

6.5 Universal Gate Sets

The concept of universal quantum gates is very important for building useful quantum computers. A universal gate set means that you have a set of gates for which any operation can be expressed as a sequence of these gates from the set. Based on the types of quantum gates we have seen so far, it seems like a universal gate set should satisfy certain criteria [29]:

(1) Create superposition states, for example, with the Hadamard gate or single-qubit rotation gates which take any $|\psi\rangle$ to any point around the Bloch sphere.
(2) Create entanglement with two-qubit gates or three-qubit gates, such as CNOT or CCNOT.
(3) Create states with complex amplitudes (e.g. $\frac{|0\rangle + i|1\rangle}{\sqrt{2}}$). While Hadamard, for example, can create superposition states, it contains only positive entries in its matrix, so you would need

another single-qubit gate in the gate set, e.g. the phase gate. However, something like an arbitrary rotation gate satisfies this criteria already.

One easy universal gate set you could think of is the set of all single-qubit operations and CNOT. Another is the set of all two-qubit gates. This result implies that there is no need to build quantum gates that involve a large number of qubits. Mathematically, there are many universal gate sets that can be constructed, but practically, there are a few that are easy to implement depending on the architecture. Furthermore, as we have started to see, errors in the qubits and gates may be more or less likely, depending on the gate set. The bottom line is that it is great that there are many possible gate sets that are universal, but some are more practical than others to implement.

Checkpoint Exercise

Determine and explain whether the following gate sets are universal. Can you express any unitary operation with the gates in this set?

(a) {Pauli gates, CNOT},
(b) {CNOT, Hadamard},
(c) {CCNOT, Hadamard, Phase[5]},
(d) {Rotation gates, CNOT}.

6.6   Quantum Teleportation and Beyond

Quantum teleportation is a pivotal protocol of many quantum algorithms [30,32]. The idea is to transfer the state of one qubit to another qubit which enables a way to overcome the complications imposed by the no-cloning theorem that we discussed in Chapter 4. Thus, even if we cannot create an exact copy of a quantum state, we could teleport

---

[5]The phase gate is

$$\begin{pmatrix} 1 & 0 \\ 0 & i \end{pmatrix}$$

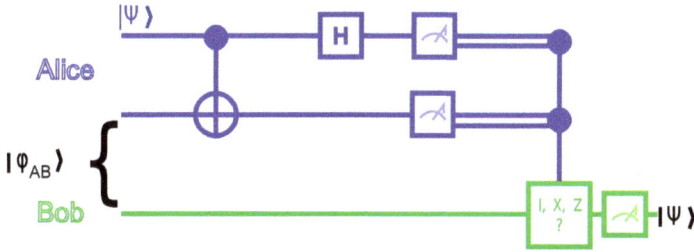

Figure 6.3. Quantum teleportation circuit for transferring a single-qubit state. Alice has two quantum registers. In one, she has prepared the single-qubit state she wants to teleport to Bob and in the second she has the qubit that she shares with Bob for the entangled Bell state. Bob has one quantum register that contains the qubit he is sharing with Alice for the entangled state, labeled $|\Phi_{AB}\rangle$.

it and destroy the original state. The teleportation here is typically over short distances, for example, transferring information between nearby qubits on a chip. This is especially useful because then we do not have to engineer all-to-all interactions between the qubits.

The original quantum teleportation protocol was proposed by Charles Bennett in 1993. The protocol involves two parties, Alice and Bob, that want to transfer some quantum state $|\psi\rangle$, as can be seen in Fig. 6.3. Let's assume that Alice has a single-qubit state

$$|\psi\rangle = \alpha |0\rangle + |\beta\rangle |1\rangle$$

For the protocol to work, there must also be a pre-shared entangled state between Alice and Bob, say, the maximally entangled Bell state,

$$|\Phi_{AB}\rangle = \frac{1}{\sqrt{2}} (|00\rangle + |11\rangle)$$

So in this case, the protocol requires three qubits, two of which are in Alice's register (one for the state she wants to transfer, and one entangled with Bob's qubit), and one in Bob's register entangled with one of Alice's qubit. Thus, the initial state of the system is

$$|\psi_0\rangle = |\psi\rangle \otimes |\Phi_{AB}\rangle$$

After this initial state preparation, Alice performs a CNOT operation between her single-qubit state $|\psi\rangle$ and her other qubit which is entangled with Bob's. This operation gives

$$|\psi_1\rangle = \text{CNOT}_A |\psi_0\rangle = \alpha |0\rangle |\Phi_{AB}\rangle + \beta |1\rangle |\Psi_{AB}\rangle$$

where $|\Psi_{AB}\rangle = \frac{1}{\sqrt{2}} (|01\rangle + |10\rangle)$, which is another maximally entangled Bell state.

Alice applies a Hadamard gate on the state she wants to transport, which puts the system in the state

$$|\psi_2\rangle = \frac{1}{\sqrt{2}} [\alpha (|0\rangle + |1\rangle) |\Phi_{AB}\rangle + \beta (|0\rangle - |1\rangle) |\Psi_{AB}\rangle]$$

Finally, we can rearrange the state by regrouping Alice's and Bob's state from above before Alice makes a measurement as follows:

$$|\psi_2\rangle = \frac{1}{2} [|00\rangle (\alpha |0\rangle + \beta |1\rangle) + |01\rangle (\alpha |1\rangle + \beta |0\rangle)$$
$$+ |10\rangle (\alpha |0\rangle - \beta |1\rangle) + |11\rangle (\alpha |1\rangle - \beta |0\rangle)]$$

Now, when Alice makes the two measurements from her register, they could be $m_1 m_2 \rightarrow 00, 01, 10, 11$. She then communicates the result to Bob through a classical channel which informs Bob's decision to make certain operations depending on the received information. When Alice makes her measurement, she has an equal probability of $\frac{1}{4}$ of getting any of the measurement results. The no-cloning theorem is satisfied in this protocol because Alice cannot recover the teleported state anymore. For example, if Alice tells him 00, then Bob just receives the state $|\psi\rangle$ without needing to do anything. If Alice tells him 10, then Bob should perform an Z-gate in order to get back to the original $|\psi\rangle$ that Alice wanted to send. You will work out what operations he needs to do in the coming exercises if Alice sends the other measurement results.

Some sanity checks about the protocol:

(1) After Alice makes her two-qubit measurement, she can no longer recover the teleported state, so the No-cloning theorem is satisfied.
(2) There is no information transfer *faster* than the speed of light since Alice has to tell Bob her measurement results over a classical channel.

So far, the protocol may not seem very intuitive since the choices of the operations done in between is unclear. In the initial state, Alice and Bob each have one bit from a pre-shared Bell state, and the entanglement here is serving as a way to connect or correlate those two bits of information. But now, Alice also holds this other qubit state that she wants to send to Bob. When Alice performs the CNOT operation with the single-qubit state and her shared qubit with Bob, she is already sharing information with Bob about the state but not in a readable way, so more operations are needed to make it clearer for Bob. So, then, performing the Hadamard gate on the state basically randomizes the outcome of her measurement in a uniform way so that the state $|\psi\rangle$ can be extracted as a result of any of the measurements Alice makes in the $\{|00\rangle, |01\rangle, |10\rangle, |11\rangle\}$ basis.

Checkpoint Exercise

1. Go through the steps of deriving the state $|\psi_2\rangle$ that is presented at the end of the section. *Hint*: Follow the quantum circuit carefully through each step.
2. What operations should Bob perform if Alice tells him that

   (a) $m_1 m_2 = 01$,
   (b) $m_1 m_2 = 10$,
   (c) $m_1 m_2 = 11$.

Homework 6

1. Distinguish two Bell states with a single shot measurement: First, generate the two Bell states called singlet, $|S\rangle$, and triplet 0, $|T0\rangle$, in Qiskit:

$$|S\rangle = \frac{1}{\sqrt{2}} (|01\rangle - |10\rangle)$$

$$|T0\rangle = \frac{1}{\sqrt{2}} (|01\rangle + |10\rangle)$$

There many potential gates you can use to generate these states, but for now, use only the Pauli gates, Hadamard and CNOT: $\{X, Y, Z, H, \text{CNOT}\}$. Here is some code that will be helpful to start:

1. Necessary imports:

```
import numpy as np
from math import *
from qiskit import *
from copy import deepcopy
```

2. Generating the state:

```
#Preparing state:
q = QuantumRegister(2)
c = ClassicalRegister(2)
circ = QuantumCircuit(q,c)
... #Flip second qubit
... #Superposes first qubit
circ.cx(q[0],q[1]) #perform CNOT gate
state = Statevector(circ)
print(state)
```

Remember this is a two-qubit state, so we cannot actually visualize the full state on the Bloch sphere as we could for a single-qubit state.

If you just measured these two Bell states in $\{|00\rangle, |01\rangle, |10\rangle, |11\rangle\}$, you would obtain equal probabilities of 0.5 for $|01\rangle$ and $|10\rangle$ for both states and could not distinguish which one you started with. Devise a way to discriminate in a single shot manner if you are measuring $|S\rangle$ or $|T0\rangle$.

Hint: Think about what rotation you can perform on the unknown state before measuring so that the measurement results help you to distinguish the states.

2. Eavesdropping during Quantum teleportation:

Suppose that Eve intercepts the teleportation of Alice's state to Bob in the following cases and show/explain whether Alice and Bob can detect that she is eavesdropping:

(a) right after Alice performs the CNOT operation between the state she wants to transfer and the shared Bell state with Bob,

(b) before Alice makes a measurement of $|\psi_2\rangle$,

(c) after Alice makes the measurement and Eve intercepts during the classical communication of the result of $m_1 m_2$.

Hint: Eve's interception is essentially making a measurement on whatever the quantum state is at the point where she eavesdrops. To determine what happens, carry out the protocol to completion accounting for the eavesdropping.

Chapter 7

Experimental Implementation

So far, we have covered the basic theoretical concepts needed to understand quantum information. In this chapter, we overview some popular state-of-the-art realizations of qubits. In particular, we focus on spin qubits and superconducting qubits. We conceptually discuss the design, implementation, performance and limitations.

7.1 Classical Computing

A fact that we may not appreciate as much as we should is that we trust our desktop PC and MAC computers to accurately implement any arbitrary computer algorithm. This is because these devices are said to be Turing-complete or computationally universal. In fact, it is not at all obvious if a given hardware can perform any algorithm that we abstract. This relies on a fine relationship between hardware building blocks and computational space.

Alan Turing was an English mathematician who formalized the concept of algorithms and computation for general-purpose computers. He developed a mathematical model which would describe an abstract machine. His theory was used as a guideline when developing classical computing hardware for data storage and manipulation based on rule sets. A fun fact is that if the hardware is a universal Turing machine, then it should be able to simulate any Turing machine. From a hardware perspective, this has an interesting consequence. Let's say we build a computer out of a certain component, e.g. light bulb, first, we need to show this is a universal computer implying that by having the lights "on" and "off," we can implement any calculation including anything our iPhones can do. This depends on the underlying mechanism that creates the notion of a "bit," in this example the lights turn on and off, and we can tell by looking at it when we "read out" the results. If we are successful in achieving a universal system, then we can be confident we can compete with any other computer made out of other components, not just light bulbs. For example, imagine someone comes up with a computer made out of relays and switches. If we have shown that our light bulb computer is universal, Turing tells us we can simulate any algorithm that the relay computer can. This is important because it gives us confidence in terms of universality. This means that if we stick to these rules for universality, everyone on the planet will get the same results when they run a program on both a light bulb and a relay computer (or any hardware that satisfies the universality rule sets). However, there could be certain advantages to using one hardware over another one. For instance, one could be faster, cheaper, more reliable (less faulty elements as it ages), smaller and lighter weight, more scalable, etc. These considerations also enter into designing qubit platforms, as we will discuss further in this chapter.

In classical computers, the hardware consists of transistors, diodes, switches, and other circuit elements. Their primary task is to perform basic operations such as AND, OR, NAND and Toffoli gates which define a set of rules. These rules are the building blocks of the computer and allow us to explore the computational space. Mathematically, we could define an algorithm with "actions" which translate to rule sets that the hardware is capable of carrying out. This is usually done through a software called "compiler" in classical computers. The rule sets understand both the operation and details of the hardware. Depending on the hardware, the sets could be based on reversible gates like Toffoli gates or irreversible ones like AND. In classical computing, the NAND gate alone provides a path to universal computing meaning any algorithm can be implemented by an architecture, perhaps complex, of only NAND components.

7.2 Universal Quantum Computing: DiVincenzo Criteria

As we have learned, quantum computing is based on reversible unitary operations. It is interesting to ask what quantum gate sets would provide universal quantum computing. Basically, we are interested in knowing what set of gates can express any unitary operation with a finite number of gates. It is not hard to see that we need more than single-qubit gates to do this. This is because single-qubit gate rotations are distinct from entangling gates. So, perhaps a combination of both would work. This has been a topic of research in the past two decades and was briefly discussed in Chapter 6. One example of a universal gate set is the Clifford set which consists of three gates (CNOT, H, S) plus T gate.

Similar to classical computation, once a set of universal gates have hardware realization, in principle, we can scale up complex architectures to implement various quantum algorithms like Shor's factoring algorithm and Grover's search algorithm. The ideal criteria is a physical system that has two quantum levels (implying that their physics is governed by the Schrödinger equation), represented by our single-qubit Bloch sphere. We will need physical ways to control its state using experimental tools such as a microwave signal generator to send microwave pulses (or laser pulses). We also need to create many of these two-level systems and make them interact through two-qubit gates to create entanglement.

Let's consider that our qubit is the spin state of a single electron. Spin is inherently a two-level system: spin-up and spin-down. In most cases, these levels are degenerate (have the same energy) until a magnetic field is applied. The value of magnetic field is proportional to the separation of the two spin levels. By tuning this energy, we can choose the frequency that could excite spin-down to spin-up and operate spin as a single qubit. Isolating an electron to manipulate its spin requires a cryogenic temperature. Typically, these measurements are done in tens of milliKelvin temperatures. For reference $1\,\mathrm{K}$ is $-273.15°\mathrm{C}$. Since the microwave frequencies (MHz–GHz) are easy to generate and can be made with very low noise, most spin qubits are operated in the microwave regime by adjusting the magnetic field to set the levels in this energy regime. As we can see, there are many considerations for choosing a platform for qubits which include the degree of stabilization, control, operation, scaling, and readout that needs to be considered. Another successful qubit platform that has risen to fame is based on superconducting circuits. Most quantum hardware companies are using a variant of superconducting qubits, each with its advantages and challenges. Big tech companies such as Google, IBM, and Amazon now have superconducting processors with hundreds of qubits and plan for thousands and more in the next few years. This is fast progress if we recall that the first realization of the superconducting qubit happened only in 1999.

Another difference among various hardware platforms (e.g. spin or superconducting) is the way each platform's quantum nature allows for certain gate sets. For example, we discussed CNOT as a primary two-qubit gate, but it is very difficult to physically implement it on certain platforms. For example, in optically based systems, it is easier to create a controlled phase gate and not CNOT. The good news is that if we have access to universal gates, then we can generate any other gate. This is usually done as part of transpiling (similar to compiling in classical hardware) to convert our algorithm from a programming language into an equivalent natural gate set of our platform.

To unify all the requirements of realization of a qubit independent of the specific platform, theoretician David DiVincenzo set several criteria necessary for constructing a quantum computer in 2000. These criteria are listed as follows:

- well-defined scalable qubit array,
- the ability to initialize the qubits into a well-defined state such as "000...",
- long enough coherence times to perform computation without losing information,
- a universal set of gate operations,
- the ability to measure any single qubit in the qubit array.

These criteria, known as DiVincenzo criteria, need to be satisfied by experimental setups to show their potential for implementing quantum algorithms, such as Grover's search algorithm or Shor's factorization algorithm.

While it may come as a surprise, there are more than 30 quantum two-level systems that can be used as qubits. Each of these platforms has their advantages and challenges. Some are being heavily researched and it is hard to say which platform will be the winner. For example, while superconducting qubits are frontrunners for solid-state systems and have the most coherent qubits, they need to be measured at milliKelvin (mK) temperatures reached by cryogenic systems called dilution refrigeration units. Typically, these qubits are about 1 mm in size which could pose challenges for scaling. In addition, controlling many such qubits could require connecting millions of cables from room temperature to milli-Kelvin dilution refrigerators which is neither easy nor scalable. On the other hand, spin qubits are small, with sizes of hundreds of nanometers, however, this proximity causes crosstalk and decoherence. In the following sections, we discuss the basics of spin-qubit and superconducting-qubit platforms and their corresponding pros and cons.

7.3 Spin Qubits

Spin qubits are typically created using nanofabrication of metallic gates. Imagine a sheet of electrons in 2D; by creating a pattern of choice and applying a negative voltage, one can remove most electrons in the sheet down to a single one. Figure 7.1 shows a schematic diagram of such a setup. Historically, the sheet of electrons can be fabricated into thin films. Researchers have developed techniques to count individual electrons using charge sensing [33]. When the last electron is trapped, we consider it a charge qubit and by applying a magnetic field, we can form the spin qubit.

The Hamiltonian, which gives the total energy of the system, is typically used to study the dynamics of qubits. In our treatment, we are thinking of the qubit as a system with two discrete energy levels, so the Hamiltonian can be expressed as a matrix. As such, we can use our linear algebra knowledge to calculate the eigenvalues and the eigenvectors.

For a spin in a magnetic field, the Hamiltonian describes the coupling between the spin and the magnetic field. For an electron, the most general form is

$$H = \frac{e}{m_e}\hat{S} \cdot \vec{B} \tag{7.1}$$

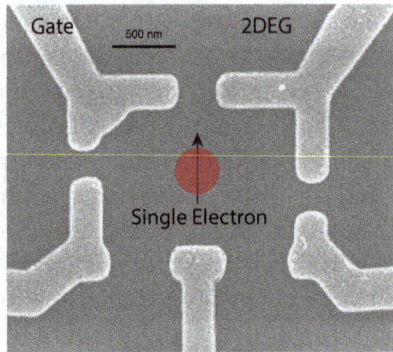

Figure 7.1. Example implementation of spin qubits. The quantum dot is defined in the two-dimensional electron gas (2DEG). The 2DEG is formed when electrons are confined within the interface of two different materials, so their motion is two-dimensional while being tightly confined in the third direction. The electron spin (denoted pictorially in orange) is confined by the surface gate electrodes. The spin is typically controlled using magnetic fields.

Recall that $\cdot$ here represents the dot product between two vector quantities. If we assume the magnetic field is oriented in the z-direction, we have

$$H_0 = \frac{e}{m_e} B_z \hat{S}_z \tag{7.2}$$

where B_z is the strength of the magnetic field in $\hat{z}$ direction, and e and m_e are the charge and mass of the electron, respectively. $\hat{S}_z = \frac{\hbar}{2} Z$ is the spin operator, which is proportional to the Pauli Z operator. So, the Hamiltonian can be re-expressed as

$$H_0 = \mu_B B_z Z = \begin{pmatrix} 1 & 0 \\ 0 & -1 \end{pmatrix} \tag{7.3}$$

where $\mu_B = \frac{e\hbar}{2m_e}$ is the Bohr magneton, a fundamental constant with units of the dipole moment. So, now, we could read off the eigenvalues, the diagonal elements of this matrix, $\mu_B B_z$ and $-\mu_B B_z$, which are the energies of the two levels. For the preceding derivation, we will redefine the value of the energy levels to be $\frac{\epsilon}{2} = \mu_B B_z$.

Now, suppose that we applied a magnetic field not only in $\hat{z}$ but also in $\hat{x}$ and $\hat{y}$. Now, the magnetic field components in $\hat{x}$ and $\hat{y}$ couple to $\hat{S}_x = \frac{\hbar}{2}X$ and $\hat{S}_y = \frac{\hbar}{2}Y$, respectively. Now, the full Hamiltonian will be

$$H = \frac{\epsilon}{2}Z + \frac{\Delta}{2}X + \frac{\tilde{\Delta}}{2}Y \tag{7.4}$$

where we have redefined the constants associated with the $\hat{x}$- and $\hat{y}$-components to be $\frac{\Delta}{2} = \mu_B B_x$ and $\frac{\tilde{\Delta}}{2} = \mu_B B_y$. These definitions have made the solution to this Hamiltonian more general. For example, now, the solution does not just have to apply to spins in a magnetic field, but it could also apply to an electric dipole coupling to an electric field ($H = -\vec{d}_e \cdot \vec{E}$), etc.

Now, let's look at it in matrix form and solve the eigenvalue problem:

$$H = \frac{1}{2}\begin{pmatrix} \epsilon & \Delta - i\tilde{\Delta} \\ \Delta + i\tilde{\Delta} & -\epsilon \end{pmatrix} \tag{7.5}$$

You will go through the derivation of the results in the checkpoint exercise. For now, the final solution to the characteristic equation is that the eigenvalues are

$$E_\pm = \pm\frac{1}{2}\sqrt{\epsilon^2 + \Delta^2 + \tilde{\Delta}^2} \tag{7.6}$$

$$\equiv \pm\frac{1}{2}\hbar\omega_q \tag{7.7}$$

where $\hbar\omega_q = \sqrt{\epsilon^2 + \Delta^2 + \tilde{\Delta}^2}$ and represents the new qubit energies. Instead of immediately saying we are in the $\{|0\rangle, |1\rangle\}$ basis or $\{|\uparrow\rangle, |\downarrow\rangle\}$, we can keep it more general and say we are in some starting basis, $\{|\varphi_+\rangle, |\varphi_-\rangle\}$. We still have the visual representation of the Bloch sphere, but now, we relabel the axes as shown in Fig. 7.2. The angular components can be defined as

$$\tan\theta = \frac{\sqrt{\Delta^2 + \tilde{\Delta}^2}}{\epsilon} \tag{7.8}$$

$$\tan\varphi = \frac{\tilde{\Delta}}{\Delta} \tag{7.9}$$

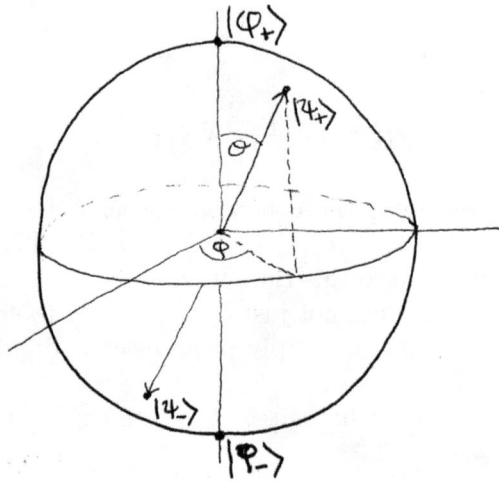

Figure 7.2. Eigensystem of a general two-level quantum system on the Bloch sphere. $|\varphi_\pm\rangle$ are the eigenstates of the pure-diagonal Hamiltonian (e.g. the spin Hamiltonian (7.2)), while $|\psi_\pm\rangle$ are the eigenstates of the general Hamiltonian with off-diagonal elements.

The eigenvectors can be written as

$$|\psi_+\rangle = \cos\frac{\theta}{2}\,|\varphi_+\rangle + e^{i\varphi}\sin\frac{\theta}{2}\,|\varphi_-\rangle \qquad (7.10)$$

$$|\psi_-\rangle = \sin\frac{\theta}{2}\,|\varphi_+\rangle + e^{i\varphi}\cos\frac{\theta}{2}\,|\varphi_-\rangle \qquad (7.11)$$

Checkpoint Exercises

1. Starting with the characteristic equation for the general Hamiltonian, find the eigenvalues:

$$\det(H - IE_\pm) = \det\begin{pmatrix} \epsilon - 2E_\pm & \Delta - i\tilde{\Delta} \\ \Delta + i\tilde{\Delta} & -\epsilon - 2E_\pm \end{pmatrix} = 0 \qquad (7.12)$$

2. Using geometric arguments, show why the angular components of the Bloch sphere could be defined as

$$\tan\theta = \frac{\sqrt{\Delta^2 + \tilde{\Delta}^2}}{\epsilon}$$

$$\tan\varphi = \frac{\tilde{\Delta}}{\Delta}$$

Hint: Recall that in a 2D plane for a vector with x-, y-components, we can define the angle it makes with the x-axis as $\tan \theta = \frac{y}{x}$. In the 3D case here, we have one 2D plane that defines the equatorial plane of the sphere and a 2D plane perpendicular to that.

7.3.1   Controlling Spins

In the previous section, we were treating the Hamiltonian in more generality. Now, let's make some useful mapping between the general case and the specific case of electron spin:

$$|\varphi_+\rangle \longrightarrow |0\rangle \tag{7.13}$$

$$|\varphi_+\rangle \longrightarrow |1\rangle \tag{7.14}$$

Let's now look more closely how the magnetic field is controlling the electron spin:

1. Considering only the z-component of the magnetic field (no off-diagonal components of the Hamiltonian), the spin state can be expressed as a superposition of $|0\rangle$ and $|1\rangle$. Let's consider how the state would evolve in time. Though the full derivation is outside the scope of the book, it turns out that if the spin initially starts out in some arbitrary state $|\psi_0\rangle = \alpha |0\rangle + \beta |1\rangle$, it will pick up a phase so that the state at a later time would become

$$|\psi(t)\rangle = \alpha e^{-i\frac{\mu_B B_z}{\hbar}t} |0\rangle + e^{i\frac{\mu_B B_z}{\hbar}t} |1\rangle \tag{7.15}$$

Factoring out $e^{-i\frac{\mu_B B_z}{\hbar}t}$, we have a state that is rotating in time, with a frequency $\omega_0 = 2\frac{\mu_B B_z}{\hbar} = \frac{eB_z}{m_e}$, which is known as the cyclotron frequency of a classical electron in a magnetic field. At time t, the azimuthal angle has changed by $\Delta\varphi = 2\frac{\mu_B B_z}{\hbar}t = \omega_0 t$.

Recall the z-axis rotation matrix, $\mathbf{R}_Z(\varphi)$ from Chapter 4. In fact, we can see that by acting $\mathbf{R}_Z\left(2\frac{\mu_B B_z}{\hbar}t\right)$ on the initial state, $|\psi_0\rangle$, we get the final state, $|\psi(t)\rangle$:

$$\mathbf{R}_Z\left(2\frac{\mu_B B_z}{\hbar}t\right)|\psi_0\rangle = \begin{pmatrix} e^{-\frac{\mu_B B_z}{\hbar}t} & 0 \\ 0 & e^{-\frac{\mu_B B_z}{\hbar}t} \end{pmatrix}\begin{pmatrix}\alpha \\ \beta\end{pmatrix} = \begin{pmatrix}\alpha e^{-\frac{\mu_B B_z}{\hbar}t} \\ \beta e^{\frac{\mu_B B_z}{\hbar}t}\end{pmatrix} \tag{7.16}$$

So, physically, having the magnetic field on in the z-direction for some amount of time rotates the state about the z-axis. The amount it rotates can be controlled by the amount of time we wait, and the strength of the magnetic field, which can be used to set the rotation frequency faster or slower. For example, if we wanted to rotate the state by $\Delta\varphi = \pi$ radians, then

$$\Delta\varphi = \omega\Delta t = \pi \implies \Delta t = \frac{\pi}{\omega}.$$

By making the frequency faster, we could rotate faster or vice versa.

2. With the off-diagonal components, the original basis, $\{|0\rangle, |1\rangle\}$ would be rotated to a new basis defined by $\{|\psi_+\rangle, |\psi_-\rangle\}$, as shown in Fig. 7.2. This new basis would take the Hamiltonian originally written in $\{|0\rangle, |1\rangle\}$ and diagonalize it so that the new energies are $E_\pm$, as we found in the previous section. Pictorially, this would mean the spin is rotating around a new axis, implying rotation of θ. So, now, if we wanted to prepare a state and turn on the magnetic field components for some amount of time, we can rotate the spin around the new axis. Then, we would get an arbitrary rotation gate, $\mathbf{R}_n(\Delta\eta)$, where n denotes the new axis and $\Delta\eta$ is the rotation angle.

One of the main issues with the static magnetic field method we have discussed thus far is that the strength of the field in the z-direction sets the rotation frequency of the spin, and realistically, this frequency ends up being much faster than desired. What if we want to keep the energy level splitting between $|0\rangle$ and $|1\rangle$ while also controlling the rate at which the qubit rotates between states? To achieve this, the answer ends up being to have a large static field in $\hat{z}$-direction, which forms the two energy levels we need, and a small time-dependent field in the $\hat{x}$- or $\hat{y}$-direction to control how fast we transition between states.

For example, let's take a magnetic field $\vec{B} = B_z\hat{z} + B_1\cos(\omega_D t)\hat{x}$. The key is to tune the frequency of the time-dependent field to be *on-resonance* so that $\omega_D = \omega_0$. This means that the field will induce a transition between the energy levels which will mix the energy levels (we will have off-diagonal components in the Hamiltonian). Similarly, when we diagonalize, we will be oriented along a new axis (essentially a $\Delta\theta$ rotation). However, now, we have fine control of the amount of rotation, which is set by the frequency $\omega_1 = \frac{\mu_B B_1}{\hbar} = \frac{eB_1}{2m}$ where B_1 was the amplitude of the time-dependent magnetic field. The calculation details are beyond the scope of the book, but if we were to start in some arbitrary location on the Bloch sphere, the final state after waiting some time is

$$|\psi(t)\rangle = \cos\frac{\omega_1 t}{2}\,|0\rangle + e^{i(\omega_0 t + \pi)}\sin\frac{\omega_1 t}{2}\,|1\rangle$$

So, we can see that azimuthal rotation, $\Delta\varphi = \omega_0 t + \pi$, is still controlled by ω_0, while the $\Delta\theta = \omega_1 t$ is controlled by the slower frequency. This discussion is the basis of nuclear magnetic resonance (NMR) which is used for magnetic resonance imaging (MRI), studying the structure of compounds, and now as a qubit control technique [34–36].

7.4 Superconducting Qubits

Superconducting circuits are one of the most promising qubit systems, currently scaling to hundreds of qubits with the potential for quantum error correction and applications in quantum simulations. In particular, IBM and Google have been pursuing this platform and have built these larger-scale systems.

Superconductivity was discovered more than 100 years ago when scientists learned to make refrigerators that could reach temperatures of $-269.15°C$ or about $4.2\,K$ [37]. This is the boiling point of liquid helium. It was discovered that if you dip a piece of superconducting material like Niobium in liquid He, its resistivity will fall to zero, forming a superconducting state. One of the first materials this phenomenon was observed in was mercury [38]. It took physicists several decades to understand the microscopic and macroscopic properties of this phenomenon, which led to many new directions in condensed matter research. The first successful theory to describe superconductivity was the Bardeen–Cooper–Schrieffer theory, commonly abbreviated as BCS theory. This theory describes how electrons in a superconductor pair up into what we call a Cooper pair. This process is mediated by vibrations in the structure of the superconductor called phonons.[1] Fundamentally, nothing stops us

---

[1]Similar to how photons are the quanta of light, phonons are the quanta of vibration.

from having a room-temperature superconductor, however now that we know what ingredients are needed like electron–phonon coupling, we can look for best-case scenarios and we still find ourselves with superconducting cases only at low temperatures (<50 K).

How do we make a qubit out of something that has zero resistance? We must make a quick detour from superconductivity to discuss the simple harmonic oscillator or SHO. Common examples of a simple harmonic oscillator are the pendulum, the mass on a spring and the electrical LC circuit. All of these systems have equivalent expressions for their equations of motion and their energy.

7.4.1   Classical Simple Harmonic Oscillator

Let's start by looking at the model of a mass on a spring. Imagine you have a block with mass m attached to a wall by a spring so that the spring can push and pull on the block and it slides back and forth without friction. Hooke's law tells us that when displacing the spring by some amount x relative to its natural length, the mass will feel a restoring force $-kx$, where k is the spring constant and x is spring's displacement from equilibrium (equilibrium is the position when the spring is not pushing or pulling on the block). The $+x$ direction is oriented away from the wall, and the $-x$ direction is toward the wall.

Now, let's look at how we can solve this problem and go from the classical case to the quantum case.

Forces and Newton's Second Law

Plugging in $F = -kx$ into Newton's second law, $F = m \cdot a$, we get a second-order ordinary differential equation:

$$-kx(t) = m\ddot{x}(t) \tag{7.17}$$

where $\ddot{x}(t)$ is a notation physicists use for the second derivative of position x with respect to time t. The details of solving the equation are not important for understanding the physics, but if you do know how to take derivatives, you can check that the following equation solves Eq. (7.22):

$$x(t) = Ae^{i\sqrt{\frac{k}{m}}t} + Be^{-i\sqrt{\frac{k}{m}}t} \tag{7.18}$$

where A and B are just arbitrary constants, determined from the initial position and velocity of the mass. Using Euler's formula, we can rewrite Eq. (7.23) in terms of sines and cosines:

$$x(t) = A\left[\cos\left(\sqrt{\frac{k}{m}}t\right) + i\sin\left(\sqrt{\frac{k}{m}}t\right)\right]$$
$$+ B\left[\cos\left(\sqrt{\frac{k}{m}}t\right) - i\sin\left(\sqrt{\frac{k}{m}}t\right)\right] \tag{7.19}$$

A mass on a spring cannot have an imaginary position, so the right A and B must be chosen so that the imaginary terms drop out. Let's leave it as an exercise to you to figure out what A and B could be in terms of some other constants C and D:

$$x(t) = C\cos\left(\sqrt{\frac{k}{m}}t\right) + D\sin\left(\sqrt{\frac{k}{m}}t\right) \tag{7.20}$$

Now that the solution is in a more intuitive form, we can see that this block will oscillate at its natural frequency or **resonant frequency**, $\omega = \sqrt{\frac{k}{m}}$, where ω is the angular frequency. You can think of angular frequency as the angular displacement of an object (measured in radians) per unit time, so the units are rad/s. So, to convert to frequency, which has units of $1/s$, we divide the angular frequency by 2π radians, $f = \frac{\omega}{2\pi}$.

Energy Perspective

Now, let's look at the same problem from a different perspective. Let us think about the kinetic energy, which is associated with the motion of the mass, and the potential energy stored in the spring. The expression for the kinetic energy of the mass on the spring is

$$KE = \frac{p^2}{2m} \tag{7.21}$$

where $p = m \cdot v$ is the momentum of the mass.

The potential energy is

$$PE = \frac{kx^2}{2} \tag{7.22}$$

Again, k is the spring constant and x is the displacement of the spring from equilibrium.

The potential energy is maximal when the spring is completely stretched out or completely compressed.[2] At the moment before the spring moves in the opposing direction, the block is stationary and has zero momentum, so its kinetic energy is 0. The kinetic energy will be maximal when the block moves through the equilibrium point of the spring, where the spring will not resist the block's motion. At this point, the potential energy of the spring is 0. So, there is this periodic energy conversion where the potential energy of the spring gives kinetic energy to the mass and vice versa.

7.4.2   Quantum Simple Harmonic Oscillator

Now, let's solve the same problem, but with quantum mechanics. The energy perspective is more easily translatable to the quantum problem than thinking about forces. So, now our mass is a quantum object whose motion is constrained by the potential energy of the spring: $\frac{k\hat{x}^2}{2}$. The Hamiltonian of a particle confined to this potential is

$$\hat{H} = \frac{\hat{p}^2}{2m} + \frac{1}{2}k\hat{x}^2 \qquad (7.23)$$

Note that we have switched from the classical position, x, and momentum, p, to position and momentum operators $\hat{x}$ and $\hat{p}$, respectively. While the full treatment of this problem is outside the scope of the book, conceptually we can understand that the position and momentum operators are continuous, so we would have an infinite-dimensional matrix to represent them. We want to solve for the eigenvalues and eigenvectors of the Hamiltonian. This will tell us the energies we expect to measure and the state of the mass. The eigenvalue problem can be expressed as we learned in Chapter 2:

$$\hat{H}\ket{\psi} = E\ket{\psi} \qquad (7.24)$$

---

[2]Of course, in reality, you don't want to stretch or compress the spring so much that you change its starting natural length because then Hooke's law would not apply.

Skipping the detailed treatment of the problem here, we have the eigenvalues of the system [26]:

$$E_n = \hbar\omega \left(n + \frac{1}{2} \right) \tag{7.25}$$

where ω is the same angular frequency as the classical version of the problem:

$$\omega = \sqrt{\frac{k}{m}} \tag{7.26}$$

$|\psi_n\rangle$ or $|n\rangle$ are the orthonormal eigenstates of the Hamiltonian. Here, n is any positive integer from 0 to infinity:

$$n = 0, 1, 2, 3, 4, ...$$

A representation of this result is shown in Fig. 7.3.

Let's first consider the first energy level which is the eigenstate $n = 0$:

$$\hat{H} |\psi_0\rangle = \frac{\hbar\omega}{2} |\psi_0\rangle \tag{7.27}$$

$|\psi_0\rangle$ is called the ground state because it is the state with the minimum possible energy. Having a non-zero minimum energy is a

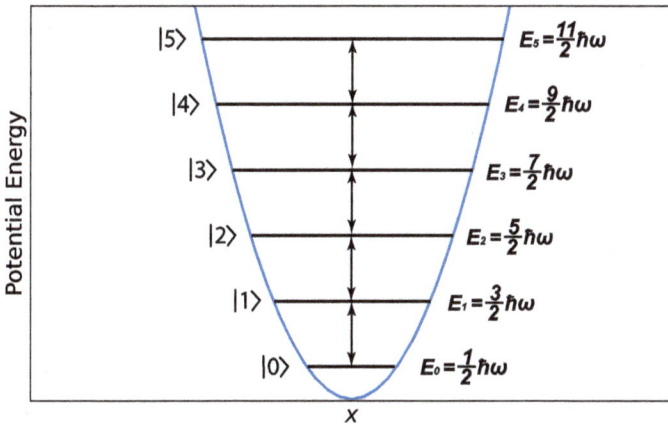

Figure 7.3. Potential energy as a function of x with the labeled quantized energy levels.

dramatic difference from the classical problem. In the classical scenario, the minimum energy is 0. This occurs when the mass is at rest in equilibrium (the momentum is 0 and the position is 0, so the total energy is 0). In the quantum version of the problem, the mass/particle cannot have zero energy; there is always a minimum energy of $\frac{\hbar\omega}{2}$. The next highest energy state is $|\psi_1\rangle$ when $n = 1$ with energy $\hbar\omega + \frac{\hbar\omega}{2}$. For $n = 2$, the energy is $2\hbar\omega + \frac{\hbar\omega}{2}$. Each successive energy level is spaced out by $\hbar\omega$. The system has "quantized" energy levels with equal spacing.

Now that we looked at the classical and quantum harmonic oscillators, we can step back and think about other harmonic systems that have identical solutions. In the example of the mass on a spring, the frequency of oscillation, ω, was determined by the constants k and m. For other SHOs, we have different constants depending on the problem. A simple pendulum's equations of motion and energy are almost identical to the mass on the spring except instead of m and k, you have g and l, where l is the length of the pendulum and g is the gravitational force on the pendulum. The angular frequency of the simple pendulum is $\sqrt{\frac{l}{g}}$.

Since the discovery of superconducting circuits like the harmonic oscillator above, researchers have shown that we can make ideal capacitors and inductors from superconductors. Capacitance is how much electric charge we can store on a conducting metal. Capacitance results from having metals or superconductors being close to another metal or superconductor and depends on the geometry of the circuit. As an analogy, the capacitor is typically thought of as the "mass" in a circuit. The inductance of a conductor can be described as its tendency to oppose a change in current. Inductors are typically coils of wire wound together. When the current varies, it causes a change in the magnetic flux, so the inductance is defined as the ratio of the magnetic flux to the current. Magnetic flux is the magnetic field times the enclosed area that contains it (in this case, the coils enclose some area). The value of the inductance depends not only on the geometry (number of windings, area enclosed) but also on the inherent properties of the metal. Keeping on with the analogy, the inductor can be considered as the "spring" in the circuit. A circuit that has capacitance, C, and inductance, L, can resonate. This means there is a resonance frequency at which charge and flux bounce back and forth. The LC circuit is yet another version of the simple harmonic oscillator.

One thing we have not included in the analysis is dissipation. For example, let's consider a pendulum. When we lift and release a pendulum at some height, we expect that it will travel the same distance to the opposite side and come back. Of course, in real life, the initial height slowly decreases until it comes to a full stop due to dissipation in the form of friction. A good pendulum could go many cycles before it slows down and bad ones only a few. In superconducting circuits, we can have resonators with millions of cycles before slowing down due to the low-loss nature of superconductors.

LC oscillator

Let's now consider a harmonic oscillator from an LC circuit. LC circuits are circuits composed of a capacitor and an inductor. Energy builds up on a capacitor with capacitance C as it accumulates stationary charge denoted by the variable Q. Energy builds up in an inductor with inductance L when current flowing through it generates magnetic flux denoted by the variable ϕ. The charge and the magnetic flux are analogous to the momentum of the mass and the position of the spring. In the same way that the energy of the mass on the spring oscillates back and forth between the potential energy of the spring and the kinetic energy of the mass, the energy in an LC circuit oscillates back and forth between the magnetic flux through the inductor and the electric charge on the capacitor. The "equations of motion" of the LC circuit are mathematically equivalent to the mass on a spring with the following substitutions:

$$x \longrightarrow \phi \tag{7.28}$$

$$k \longrightarrow \frac{1}{L} \tag{7.29}$$

$$p \longrightarrow Q \tag{7.30}$$

$$m \longrightarrow C \tag{7.31}$$

$$\tag{7.32}$$

The same math we used to solve the classical and the quantum SHO map over to the LC circuit problem. At room temperature, an LC circuit behaves like a classical mass on a spring. When the LC circuit is cooled down to superconducting temperatures, the Cooper pairs have condensed into a single wavefunction with observables: $\hat{\phi}$ and $\hat{Q}$. We then arrive at the Hamiltonian:

$$\hat{H} = \frac{\hat{Q}^2}{2C} + \frac{\hat{\phi}^2}{2L} \tag{7.33}$$

The Hamiltonian in Eq. (7.33) is equivalent to Hamiltonian in Eq. (7.23), but now, the resonant frequency is $\omega = \sqrt{\frac{1}{LC}}$ instead of $\omega = \sqrt{\frac{k}{m}}$.

Now, how do we make this superconducting LC circuit into a qubit? At first glance, we could map the ground state $|\psi_0\rangle$ to $|0\rangle$ and the first excited state to $|\psi_1\rangle$ to $|1\rangle$, however there is a huge flaw with this. Imagine we measure the LC circuit to determine the state that it is in: We will have a spectrum of photons, each at a different energy. If the LC is in the ground state, it will absorb the photon with energy $\hbar\omega$ and jump up to the first excited state. If the circuit is already in the first excited state, and you repeat the experiment, it will also absorb the photon with energy $\hbar\omega$ and jump up to the second excited state. No matter what state you are in, it will always absorb the photon with energy $\hbar\omega$ because all of the states are evenly spaced out by $\hbar\omega$. There is no way to tell if your superconducting resonator is in $|\psi_0\rangle$, $|\psi_1\rangle$ or $|\psi_n\rangle$.

So, we need a way to make the levels anharmonic (not evenly spaced) so that we could isolate two energy levels that we can call the ground and excited states of the qubit, as can be seen in Fig. 7.4. One circuit element that solves this issue is called the "Josephson junction." A Josephson junction (JJ) consists of a thin layer of insulating material sandwiched between superconducting metal. The JJ is typically less than 2 nm thick. The Cooper pairs in the superconductor can no longer flow freely through this barrier, but they can tunnel through the barrier. The junction adds a nonlinear component to the Hamiltonian, $E_J \cos\hat{\phi}$:

$$\hat{H} = \frac{\hat{\phi}^2}{2L} + E_J \cos\hat{\phi} \tag{7.34}$$

The new term adds anharmonicity to the LC circuit. It staggers out the energy levels so that the energy transition from the ground

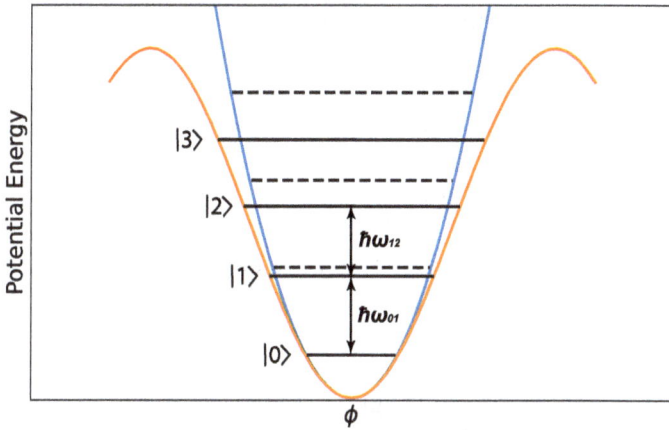

Figure 7.4. Representation of the potential energy going from harmonic to anharmonic, which causes the energy levels to stagger out and thus allows us to isolate two energy levels.

state to the first excited state is no longer exactly the same as the energy transition from the first state to the second excited state. This effect can be visualized in Fig. 7.4. Now, we can go back to our original idea to map the ground state to $|0\rangle$ and the first excited state to $|1\rangle$.

We know superconducting circuits have to be cold, but how cold is cold? The answer is that these circuits work below the critical temperature of the superconductor. For aluminum, this is 1.2 K, and for niobium, it is 9 K. Let's convert temperature to frequency, so we can make a direct comparison. The thermal energy of a given temperature, T, is $k_B T$, where k_B is the Boltzmann constant. As discovered by Planck and Einstein, the quanta of energy is $h\nu$, where h is Planck's constant and ν is the frequency of the photon. Equating these two relations tells us that 1 GHz is 48 mK, 2 GHz is 96 mK, 5 GHz is 240 mK, and so on. Aluminum has a critical temperature of 1.2 K, which is about 23 GHz, while niobium (critical temperature is about 9 K) is 194 GHz. Superconducting circuits are typically operated in a frequency range between 2–10 GHz. This is for two reasons: (1) existing microwave components and generators can be used to reliably control and measure the qubit in the 2–10 GHz regime; (2) we want it to be much colder than the critical temperature of the superconductor (and therefore lower frequency) to reduce unwanted thermal excitations which degrade qubit performance.[3]

A typical energy spacing for superconducting qubits is 5 GHz where we have access to pristine microwave tools and operate at the base temperature of a dilution fridge, 10 mK. Now, what does this imply about the size of our resonators? The speed of light, $c \approx 3 \times 10^8$ m/s, equals the product of the wavelength, λ, and the frequency, f: $c = \lambda \cdot f$. So, for 5 GHz, the wavelength of the light is about 60 mm. So, you can imagine that if we had a cable that long just filled with air, then it would resonate at 5 GHz. In reality, this cable would be made on a silicon or sapphire chip and would be coupled to the qubit, so the speed of light is reduced, so that the size of the on-chip resonators can be a few millimeters.[4] In one way, this is an amazing finding that we can have a fully quantum mechanical object that is millimeter scale completely governed by the Schrödinger equation. On the other hand, compared to spin qubits, we could say this is a big object and not very favorable for scaling. One million superconducting qubits could become a meter-long object, but we should remind ourselves that the first classical computers were also the size of a room.

7.5   Performance Comparisons Across Different Qubit Platforms

Some of the key performance metrics that will be reported to compare qubit platforms are the coherence time, often noted as T_2 time, and the relaxation time, often noted as T_1 time. Physically, the coherence time represents the amount of time that the qubit can retain its

---

[3]This is essentially saying that if we want to have the qubit in the ground state or $|0\rangle$, we do not want a hot thermal environment to excite the qubit to the excited state or $|1\rangle$ erroneously. A simple statistical mechanics argument shows that, for example, if we prepare the qubit in $|0\rangle$ and want it to be in that 99.99% of the time, then we should lower the temperature down to around 100 mK. So, now, we need dilution refrigerators to reach such low temperatures. In practice, the base temperature of the dilution fridge can be as low as 10 mK to reduce these unwanted thermal excitations.

[4]The speed of light is about 7.5×10^7 m/s in silicon ($4 \times$ slower than in vacuum), and about 1.5×10^8 m/s in sapphire ($2 \times$ slower than in vacuum).

superposition state. Imagine doing the following experiment: You start with the qubit in the ground state (positioned up along the z-axis of the Bloch sphere), and you apply a 90° R_x or R_y rotation to put it on the equatorial plane of the Bloch sphere. Then, you wait for a certain amount of time until the phase or superposition state completely decoheres. The amount of time that it took is essentially T_2. Now, imagine that you again start with the qubit in the ground state and you apply a 180° R_y rotation, which is also known as a π-pulse. Then, you wait for the qubit to relax or decay back to the ground state. The amount of time that it takes is the T_1 time. These timescales are measured across different qubit platforms and are used to compare what is best. The combined effect of these timescales has huge implications on how useful a platform will be for running large-scale useful and interesting algorithms.

So, let's briefly compare the performance of the two platforms we discussed, spin qubits and superconducting qubits:

1. In spin qubits, one practical advantage is that they are small (nanoscale size) and can therefore be easily scaled up. As we discussed, these qubits can be controlled using NMR and electron-spin resonance (ESR) techniques which can isolate the qubit and protect it from unwanted interactions. Another advantage is depending on the implementation, the qubits could work even at room temperature.[5] The coherence times of spin qubits using quantum dots are typically of the order of a few microseconds and not as long as state-of-the-art superconducting qubits [40].

2. In superconducting qubits, we have the advantage of working at milliKelvin temperatures with very low dissipation and thermal excitations which allows for making devices with little loss of information. Further, the energy scale of the qubit allows us to use existing microwave tools that provide low-noise operation in the 2–10 GHz regime. Electrical engineers, especially in the aerospace and telecommunications industries, have great familiarity and experience making low-noise sources, amplifiers, and

---

[5]If you are curious, spin qubits can be implemented by trapping electrons using quantum dots (which are devices made on a chip and controlled with microwave signals), but they can also be nuclear spins of a diamond, for example, called *NV* centers, which can work at room temperature and are laser-controlled [39].

detectors in this frequency regime. Some of the state-of-the-art relaxation times and coherence times achieved in these qubits (at the time this edition of the book was published) are 1 ms and 1.5 ms, respectively [41]. By studying different ways to engineer these qubits, research has come a long way from the original superconducting qubits which only had a coherence time of nanoseconds!

There are heaps of other performance metrics that matter including the speed at which gates can be performed, the fidelity or accuracy of those gates, the fidelity of preparing a particular state, being able to reliably couple and control multiple qubits, and so on. We will not go into those details here but let's consider an example, to give you an idea of how delicate the problem of engineering qubits could become. Superconducting qubits have some of the fastest gates while atom-based or ion-based qubits have slower gate times. However when you divide the coherence time by the gate clock period, you could achieve more or less the same number of operations. So for a given algorithm they provide the same number of coherent gate operations. Right now, it is not clear if they outperform each other because qubit systems are multifaceted. It is not just one metric that decides which platform is better. The higher the time a qubit can maintain its quantum state is clearly important, but so is the speed of operations and the ease of being able to entangle qubits.

Conclusion

You have arrived to end of this book and your head might be filled with visions of a quantum future. It has been more than a century full of discoveries from the inception of the theory of quantum mechanics. We have now realized many conceptual ideas into laboratory realities. This century started with a more fascinating dream of not just observing and discovering new quantum phenomena but also trying to go beyond by very precisely controlling quantum states. Building and controlling quantum states, like qubits, may unlock solutions to longstanding problems that are known to be computationally difficult. This technology may become part of our every day life in the next century or sooner.

Quantum computing is a highly multidisciplinary topic and you could find the footprints of many science disciplines in it. From physics to chemistry, from computer science to software engineering, from microwave engineering to materials science. Everyone plays a crucial role. This book aims to provide entry-level but practical background to those who are keen on learning quantum computing, but have not gone through the rigorous quantum physics education. The book covers basic mathematical concepts as well as important thought experiments and laboratory discoveries that were key to the formation of our quantum computing platforms today. We provided the basics of linear algebra as it is necessary to understand quantum information. Using this mathematical background, we gained physical intuition by describing various physical situations of how qubits could be implemented, how they can be controlled through gates or matrices, and how measurement is tied to the eigenvalues and eigenvectors of those matrices.

While a single-qubit representation is relatively simple, we described a few examples of how crucially different quantum concepts could be from their classical counterparts. We continued by extending number of qubits to two and introducing entanglement. The EPR paradox shows how the founders of quantum mechanics had trouble accepting the concept of entanglement and non-locality. While it can be debated from a philosophical point of view, entanglement is real and defines the real power of quantum computers. We provided two examples of spin and superconductivity as physical platforms where these concepts can be employed in reality.

This book is a starter's guide, but by no means is it completely covering every subject. This strategy is used throughout the book with the goal of providing essential materials for understanding practical and widely used quantum information concepts. The state of the art will surely march forward, and independent of whichever platform will be the forerunner, the concepts described here remain the same. We hope that Qiskit and Quantum Flytrap softwares could serve as virtual labs to help with visualization of abstract concepts being developed in real laboratories.

Chapter A

Coding with Qiskit

Qiskit was initially released on March 2017 by IBM Research. It is an open-source framework for quantum computing that allows users to develop and manipulate quantum programs and run them on real quantum processors through IBMQ, simulators and even low-level simulations through OpenQasm [42]. The primary version of Qiskit uses the popular programming language, Python, but versions for JavaScript are used for OpenQasm. We will be using the primary version of Qiskit, simulating quantum circuits and understanding their behavior through the Bloch sphere.

Qiskit is made up of four different *elements*:

1. **Terra**: This allows for composing quantum circuits where you define the number of qubit registers and classical registers, which store results after measuring the circuit. There also exist visualization tools to plot measurement results.
2. **Aer**: This contains the simulators for the quantum circuits, including the *Statevector simulator*, which is what our Bloch sphere is trying to physically show, the *Qasm simulator*, which runs circuits multiple times and returns probabilistic results, and the *Unitary simulator*, which returns a matrix that represents the circuit that the user built.
3. **Ignis**: Real quantum hardware is susceptible to many different kinds of random errors or *noise*. If a circuit is run on the IBMQ processors, the results will not be 100% accurate due to the presence of noise. This element of Qiskit provides ways to

characterize and mitigate errors so that circuits can provide more accurate results even in the presence of noise.

4. **Aqua**: This element of Qiskit operates at the highest level thus far since it contains algorithms that can be used to build applications to solve problems. The main problems where quantum computing may have a real advantage are in chemistry (since molecules are inherently quantum systems), artificial intelligence, optimization and finance.

As more research is published and interest in quantum computing rises, Qiskit remains an open community, taking input from users, researchers and software developers. For information on Qiskit, read their documentation (https://qiskit.org/documentation/index.html) or look through Qiskit Learn (https://qiskit.org/learn) for tutorials to get started.

Qiskit Installation

Currently, the optimal way to install Qiskit is by using virtual Python environments. Specifically, we will use Anaconda, a free, open source distribution to simplify package management for Python which includes CLI and GUI interfaces through Anaconda Power-Shell Prompt and Anaconda Navigator, respectively. When working in a *conda environment*, all dependencies and package versions are checked automatically to ensure that there are no conflicting dependencies.

First, install latest version of Anaconda to your machine. The appropriate versions for your OS can be found at: https://docs.ana conda.com/anaconda/install/.

After installing Anaconda, open the Anaconda PowerShell Prompt and create a new environment to use Qiskit by typing the follow commands:

```
conda create -n qiskit_env python=3
```

Running this command will list several packages that will be installed and ask you whether to proceed. Type *y* when asked and then activate your environment:

```
conda activate qiskit_env
```

The next step is to install some additional packages that will be necessary to use along with Qiskit. This includes NumPy, SciPy, matplotlib and Jupyter Notebook. Type the following command to begin installation:

```
conda install numpy scipy matplotlib jupyter notebook
```

Again, after running the command, you will be asked to install several packages, so type *y* to proceed.

Now, we are ready to install Qiskit! Type:

```
pip install qiskit
```

Once the installation is complete, also install visualization tools:

```
pip install qiskit[visualization]
```

Now, you can start up Jupyter Notebook by typing:

```
jupyter notebook
```

This will open up a web page that has the files on your computer. You can create a new folder to store the notebooks where you work with Qiskit. When you open a new notebook, you will note that there is an empty block. This is where you can type your code and then press **Shift + Enter** to execute and output will appear below. A new code block is created every time you execute a previous one, but you can always go back to previous cells, edit and rerun.

Start off by importing Qiskit to ensure that the installation was successful. Executing the following code should output:

```
In [1]:  from qiskit import *
         qiskit.__qiskit_version__
Out[1]:  {'qiskit-terra': '0.14.2',
          'qiskit-aer': '0.5.2',
          'qiskit-ignis': '0.3.3',
          'qiskit-ibmq-provider': '0.7.2',
          'qiskit-aqua': '0.7.3',
          'qiskit': '0.19.6'}
```

Alternatively, you can access JupyterLab online through Anaconda Cloud service via the link: https://anaconda.cloud/.

Quantum Flytrap: Virtual Optical Table

We will use Quantum Flytrap (https://quantumflytrap.com/virtual-lab), a virtual optical table, as a resource to understand how to generate different kinds of light polarization and how they can be measured in different bases.

The default light source is a laser that provides horizontally polarized light. Using the many optical tools that are offered, we can modify the polarization and observe it in different bases. Some tools are described in the following and also shown in Fig. B.1.

1. **Flat mirror**: This is used to reflect incident light by the same angle.
2. **Beam splitter**: This splits beam of incident light into two, where some percentage of light is transmitted through the beam splitter and the rest is reflected.
3. **Polarizer**: This is the optical filter that may be oriented to only allow light of specific polarization to pass through and block light of other polarizations.
4. **Wave plate**: This is also known as *retarder* and is a device that alters the polarization of a light wave traveling through it. This is of two types:

   Half-wavelength plates: These shift the polarization direction of linearly polarized light and delays phase by 180°. This allows us to switch from horizontal to vertical or angles between 0–90°, depending on the orientation of the plate.

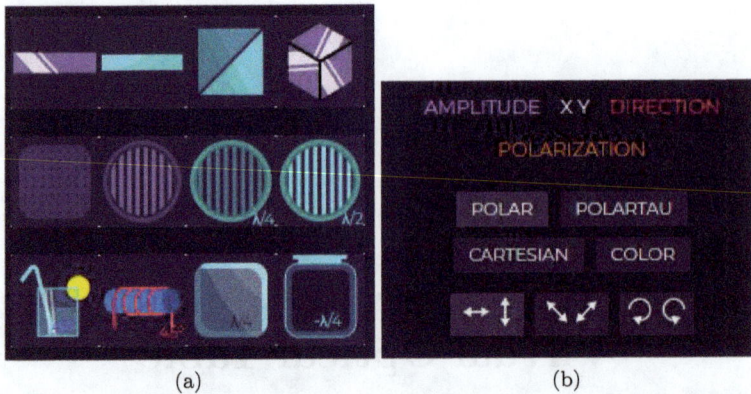

Figure B.1. (a) Optical tools in Quantum Flytrap. *First row*: mirror, beam splitters, reflective cube; *second row*: light absorber, polarizer, quarter-wavelength wave plate, and half-wavelength wave plate; *third row*: sugar solution, Faraday rotator, glass and vacuum jar for changing phase. (b) Measurement bases can be viewed when enabling "Wave" mode.

Quarter-wavelength plates: These shift the polarization direction from linear to circular and vice versa, and delay the phase by 90°.

5. **Faraday rotator**: This is a device that involves sending light through a magnetic field, which rotates the polarization state.

These tools may be dragged into the optical table and rotated as necessary. The changes in the quantum state can be tracked in different bases and are shown to the right of the screen. Let's take the default horizontal polarization and turn into vertical. We can do this using the half-wavelength plate, which delays polarization by one half of a wavelength, as shown in Fig. B.2. We must rotate the wave plate by 45° because the wave plate delays the phase at twice the angle that it's rotated at, thus achieving a 90° shift.

Similarly, we can generate circularly polarized light by using a quarter-wavelength plate, which delays polarization by one quarter of a wavelength. Once again, we rotate the wave plate at a 45° angle so that the incident line is coming through at 45° so that the quarter-wavelength plate can shift the polarization to circular. Specifically, we generate right-handed circular light, as shown in Fig. B.3.

If we wanted to generate left-handed light, then we would have to rotate the wave plate by an additional 90°.

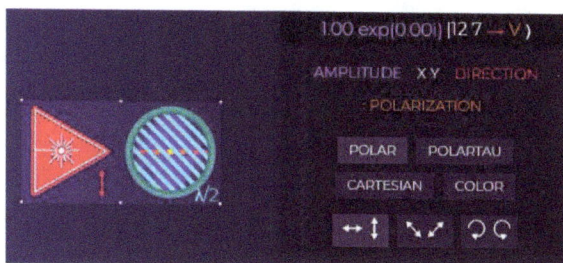

Figure B.2. Changing polarization from horizontal to vertical.

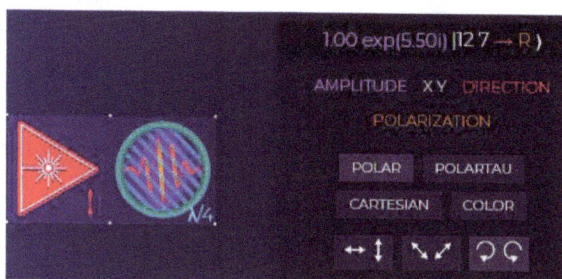

Figure B.3. Changing polarization from horizontal to circular right-handed light.

Generating Photon Superposition Using Beam Splitters

The beam splitter is a device used to split light and recombine it in certain cases. A beam splitter does not change the polarization of light, but instead splits it and the split beams move in different directions. The beam splitter is made of glass and has a dielectric coating on one side. Light that is reflected from the dielectric side picks up a π phase shift. Light reflected from the side of the glass does not pick up any phase shift. Finally, all transmitted light experiences no phase shift. This can be seen in Fig. B.4. If we think about the wave nature of light, a π phase shift means that each peak of the original light is shifted forward by one. This is important when we have two beams that later recombine, as the π phase shift in one will mean *destructive* interference!

We can also generate a superposition state by using a 50/50 beam splitter, where 50% of the light intensity is transmitted (goes through the beam splitter) and 50% is reflected. In the optical table, we drag

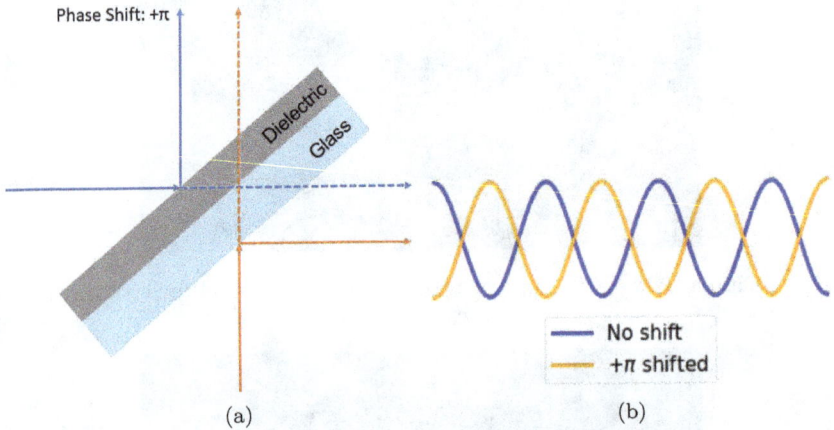

Figure B.4. (a) Light that is reflected from the dielectric side (blue arrows) is phase shifted by π rad, while light reflected from glass side (orange arrows) has no phase shift. Any light that is transmitted experiences no phase shift. (b) Visual of $+\pi$ or $180°$ phase shift for a wave.

the beam splitter and rotate it at a $45°$ angle so that the incident horizontal light reaches the beam splitter at a $45°$ angle. We also place two detectors in the path we expect the light to follow and observe which detector triggered.

In this experiment, we have the laser providing us single photons. For each photon that we send in, only one of the detectors is triggered. In this way, the photon is behaving like a particle instead of a wave. If we sent many photons in one by one and gathered the statistics for the number of times either detector was triggered, we would see that for a 50/50 beam splitter, either detector may be triggered with an equal chance (you may try this by enabling the "Loop" setting under "Waves"). This is because the photon state, which we will denote as $|P\rangle$, is in a superposition until it reaches the detectors, as shown in Fig. B.5(a):

$$|P\rangle = \frac{1}{\sqrt{2}}(|\rightarrow\rangle + |\uparrow\rangle)$$

What if we tried a beam splitter with a different splitting ratio like 60/40? We can right click on the beam splitter and change its reflectivity to 60%. Now, 60% of the incident light is reflected, and only 40% is transmitted, as shown in Fig. B.5. Once again, if we

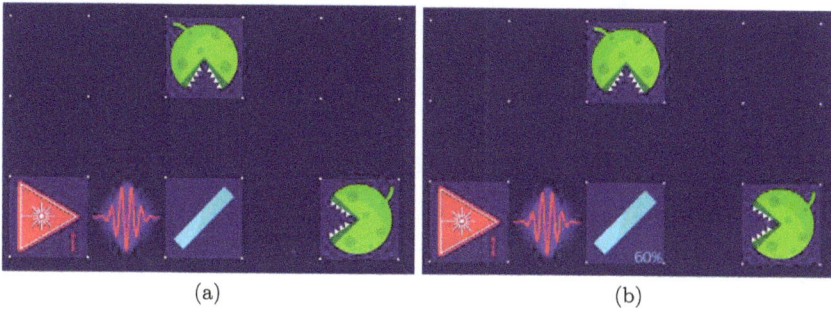

(a)   (b)

Figure B.5. (a) 50/50 Beam splitter setup to generate a superposition. Green monsters are detectors that measure the state of the light reaching them; (b) 60/40 beam splitter setup to generate a superposition.

sent many single photons one by one and gather statistics, we would see that the detector oriented along the path of the reflected photon direction will be triggered more often because we set the beam splitter to reflect more light. The photon still behaves like a particle, so it will only reach one of the detectors, but this change in the beam splitters alters the superposition so that we have a 60% chance of reaching the detector positioned along the path of the reflected photons vs. only a 40% chance of triggering the detector positioned in the path of the transmitted photons. Alternatively, if we had made the beam splitter only 40% reflective then, 60% of the photons would be transmitted, thus the detector positioned along the path of the transmitted photons has a 60% chance of being triggered.

Chapter C

Additional Math Review

This is a reference sheet for additional math content that will be necessary.

Trigonometry

Trigonometry studies right-angled triangles or right triangle and uses trigonometric functions to relate an angle of right triangles to the ratio of two of its side lengths. The hypotenuse of a right triangle can be computed by using the Pythagorean theorem: $c^2 = a^2 + b^2$.

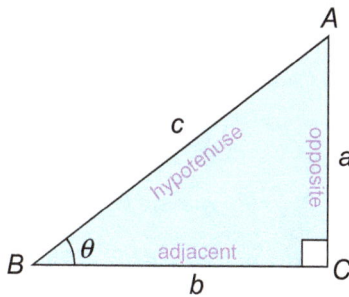

Figure C.1. Right triangle.

The trigonometric functions are known as *cosine*, *sine*, and *tangent*, and they are defined in the following manner:

$$\cos \theta = \frac{\text{adjacent}}{\text{hypotenuse}} = \frac{b}{c}$$

$$\sin \theta = \frac{\text{opposite}}{\text{hypotenuse}} = \frac{a}{c}$$

$$\tan \theta = \frac{\text{opposite}}{\text{adjacent}} = \frac{a}{b} = \frac{\sin \theta}{\cos \theta}$$

If we know the side lengths of the triangles, we may compute the angle θ[1] by performing an inverse operation on the trigonometric functions, which can be denoted as $^{-1}$ or with the prefix "arc":

$$\theta = \cos^{-1}\left(\frac{b}{c}\right) = \sin^{-1}\left(\frac{a}{c}\right) = \tan^{-1}\left(\frac{a}{b}\right)$$

$$\theta = \arccos\left(\frac{b}{c}\right) = \arcsin\left(\frac{a}{c}\right) = \arctan\left(\frac{a}{b}\right)$$

Unit Circle

A circle is a curve defined in terms of its radius r. If we plot the circle on a Cartesian plane, then any point on the circumference of the circle can be labeled as (x, y). A **unit circle** is a circle centered at the origin and has a radius of 1. The x coordinate can be described using "cos", and the y coordinate can be described using "sin":

$$x = r \cos \theta$$

$$y = r \sin \theta$$

---

[1] Angles may be written in the form of degrees, noted as $^\circ$ or in radians, noted as rad. A full circle has $360°$ or 2π radians, and we may convert from radians to degrees using the ratio:

$$\frac{360°}{2\pi} = \frac{180°}{\pi}.$$

We may also convert from degrees to radians using the reciprocal of that ratio: $\frac{\pi}{180°}$.

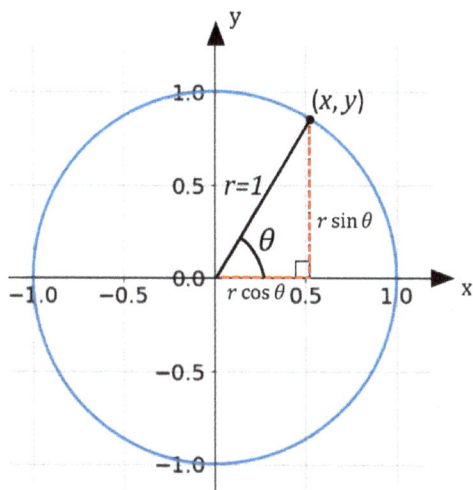

Figure C.2. Unit circle.

Using the Pythagorean theorem, we must have that

$$x^2 + y^2 = r^2 = (r\cos\theta)^2 + (r\sin\theta)^2 = r^2(\cos^2\theta + \sin^2\theta)$$

From this, we obtain the trigonometric identity:

$$\cos^2\theta + \sin^2\theta = 1.$$

There are some common angles that are often encountered for which we know the value of "cos", "sin", and "tan":

Angle, $\theta°$	Angle, θ rad	$\cos\theta$	$\sin\theta$
0°	0	1	0
30°	$\pi/6$	$\sqrt{3}/2$	$1/2$
45°	$\pi/4$	$\sqrt{2}/2$	$\sqrt{2}/2$
60°	$\pi/3$	$1/2$	$\sqrt{3}/2$
90°	$\pi/2$	0	1

The table shows the angles in the first quadrant, however, the cos, sin and tan of other angles can be related to these angles which are known as **reference angles**. For example, if we wanted to know $\cos 135°$, the reference angle can be computed as $180° - 135° = 45°$.

The only thing we have to watch out for is the sign. We can see from the unit circle that if we have $\theta > 90°$, then the x-components are negative, and since *cos* represents the x-components, then $\cos 135° = -\frac{\sqrt{2}}{2}$.

If $\theta > 180°$, say, $270°$, then the reference angle can be computed as $360° - 270° = 90°$. Again, we have to watch out for the sign, looking at the unit circle, $\theta = 270°$ corresponds to the negative y-axis. So, $\cos 270° = 0$, but $\sin 270° = -1$.

Lastly, the convention for positive angular direction is *counter-clockwise*, ↺, and that for the negative direction is *clockwise*, ↻. For example, if we have an angle that is greater than $180°$, we may also write it conveniently in the clockwise direction:

$$270° = -90°$$

$$315° = -45°$$

Rules for Exponents

An exponential includes an exponent which represents repeated multiplication of some number, which is called the base. So, we have $2^3 = 2 \cdot 2 \cdot 2$, where 2 is the base and 3 is the exponent. The rules for working with exponents are general and apply whether the exponent is a specified number or variable:

1. Zero-exponent rule states anything raised to the zero power is 1:

$$a^0 = 1$$

2. Power rule:

$$(a^m)^n = a^{m \cdot n}$$

3. Negative exponent rule:

$$a^{-n} = \frac{1}{a^n}$$

4. Product rule[2]:

$$a^m \cdot a^n = a^{m+n}$$

5. Quotient rule[3]:

$$\frac{a^m}{a^n} = a^{m-n}$$

---

[2]Note that the base of the two exponentials must be the same to apply this rule, if they are not, you cannot perform this simplification.
[3]Again, bases of the exponentials must be the same for this rule to apply.

Checkpoint Exercise Solutions

Chapter 1

Probability

Introduction

First, we look at the sample space for this probabilistic experiment. We may use a table to visually see what our sample space is:

	1	2	3	4
1	1,1	1,2	1,3	1,4
2	2,1	2,2	2,3	2,4
3	3,1	3,2	3,3	3,4
4	4,1	4,2	4,3	4,4

As we can see, we have 16 possible outcomes which are all equally likely. So the probability of each event occurring is $\frac{1}{16}$. Now, we want the probability that the sum of the two outcomes is 5, and we can see from the table that this occurs for four events: (1,4), (2,3), (3,2), (4,1). Thus,

$$\mathbb{P}(\text{sum} = 5) = \frac{4}{16} = \frac{1}{4}$$

Combinatorics

1. There are 3 places for letters and 2 places for numbers and we want *unique* license plates, implying that we don't repeat any letters or numbers. There are 26 letters in the alphabet and 10 choices for numbers from 0 to 9, so we have the following:

$$\underline{26} \times \underline{25} \times \underline{24} \times \underline{10} \times \underline{9} = 1,404,000$$

2. There are 3! ways to arrange 3 math textbooks because we have 3 possible choices for the first place, 2 for the second, and 1 for the third, so $3 \times 2 \times 1 = 3! = 6$. Similarly, there is only one way to arrange the chemistry textbook: $1! = 1$. Lastly, there are $5! = 5 \times 4 \times 3 \times 2 \times 1 = 120$ ways to arrange our beloved physics textbooks. Now, the last step is to figure out the possible ways that the textbooks can be arranged, for example, which sets of textbooks go on the left, next, and so on. This is the same as considering the number of ways to arrange 3 objects since we have 3 different subjects of textbooks here. This may be arranged in 3! ways. So, our final result is

$$3! \cdot (3! \cdot 1! \cdot 5!) = 4320$$

3. (a) Here, we are trying to arrange the players so that *Team 1* players stand together. That means that *Team 1* can be considered as one group while the players from *Team 2* can be treated as 6 distinct groups since we don't have this restriction that they all have to stand next to each other. This gives us 7 distinct groups that can be arranged in $7! = 5040$ ways. For example, if *Team 1* is positioned in the middle and they stand all together, some players from *Team 2* may stand to the left or to the right of *Team 1*. Now, we know how many ways we could arrange these distinct groups. The last step is to consider that while the players in *Team 1* stand next to each other, there are $5! = 120$ ways that the players can be arranged within the group. Our final answer is thus

$$7! \cdot 5! = 604,800$$

(b) Now, we are choosing 3 players which must all be either from *Team 1* or *Team 2*. This is a combinations problem: there are $\binom{5}{3}$ ways to choose 3 players from *Team 1* and $\binom{6}{3}$ ways to choose 3 players from *Team 2*. Since we are saying "or" and the occurrence of these events occurring is disjoint, we add the number of ways:

$$\binom{5}{3} + \binom{6}{3} = \frac{5!}{3!(5-3)!} + \frac{6!}{3!(6-3)!} = 30$$

Discrete Random Variables

1. (a) If we toss a coin twice and keep track of the number of heads that appear, the random variable S may be 0, 1, 2. This means that we may get no heads for both toss times, only one head or two heads for both times.

   (b) Each outcome of the coin is equally likely and independent. So, if we got 0 heads, then that means we got 2 tails instead and this has a probability of $\mathbb{P}(0H) = \frac{1}{2} \cdot \frac{1}{2} = \frac{1}{4}$ of occurring. If we get one head, then the other time we tossed, we got tails, but the head could appear either on the first or second toss, so the probability of getting one head in this experiment is $\mathbb{P}(1H) = 2 \cdot \frac{1}{2} \cdot \frac{1}{2} = \frac{1}{2}$. Lastly, if we got 2 heads, then

$$\mathbb{P}(2H) = \frac{1}{2} \cdot \frac{1}{2} = \frac{1}{4}$$

This experiment can be represented by the following binomial distribution:

$$\mathbb{P}(S = k) = \binom{2}{k}\left(\frac{1}{2}\right)^{k}\left(\frac{1}{2}\right)^{2-k}$$

Using our rules of exponentials, we can reduce this to

$$\mathbb{P}(S = k) = \binom{2}{k}\left(\frac{1}{2}\right)^{(2-k)+k} = \binom{2}{k}\left(\frac{1}{2}\right)^{2} = \binom{2}{k}\left(\frac{1}{4}\right)$$

2. Each free throw Eddie makes can be considered as a Bernoulli trial since there is a probability of him succeeding and failing. We are given the probability of Eddie succeeding at making the free throw: $p = 0.8$. The number of times he attempts the free throws is $n = 8$. So, we can define a random variable, say T, to keep track of the number of free throws Eddie makes successfully. T can take following values: 0, 1, 2, 3, 4, 5, 6, 7, 8. This means Eddie may

fail at all free throws and succeed at making some of them or all of them. Thus, we can write the Binomial distribution:

$$\mathbb{P}(T = k) = \binom{8}{k}(0.8)^k(0.2)^{8-k}$$

We are asked to calculate the probability that he makes at least 3 of these free throws successfully, which we represent as $\mathbb{P}(T \geq 3)$. Making at least 3 successfully means that he could have made 3, 4, 5, 6, 7 or 8 free throws successfully. So, we would need to add up $\mathbb{P}(T = 3)$, $\mathbb{P}(T = 4)$ ... $\mathbb{P}(T = 8)$. Another way to reframe the answer and cut down on so much computation is to say Eddie making at least 3 successfully is the same as if he **does not** make 0, 1, 2 free throws:

$$\mathbb{P}(T \geq 3) = 1 - \mathbb{P}(T \leq 2) = 1 - \mathbb{P}(T = 2) - \mathbb{P}(T = 1) - \mathbb{P}(T = 0)$$

Plugging these into the equation for the distribution:

$$\mathbb{P}(T \geq 3) = 1 - \binom{8}{2}(0.8)^2(0.2)^6 - \binom{8}{1}(0.8)(0.2)^7$$

$$- \binom{8}{0}(0.8)^0(0.2)^8 = 0.99877$$

So, because Eddie's probability of success is quite high, we can understand how he is able to make many free throws successfully. If we look at the distribution, we also note how the highest probabilities occur for $T \geq 3$:

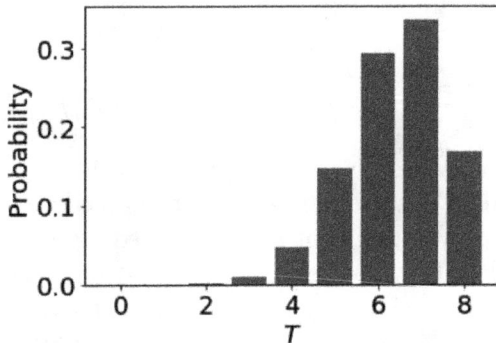

2. In this problem, we view the effectiveness of the vaccine as a Bernoulli trial. In this case, the Pfizer vaccine has 0.8 effectiveness in preventing COVID-19. We have $n = 6$ patients. If we define a random variable I which keeps track of how many patients are infected after receiving the vaccine, then I can be 0, 1, 2, 3, 4, 5, 6. This means that 0 patients may contract the virus after receiving the vaccine, some may contract it, or all may contract it. So, now, since we are keeping track of who gets infected, we need to write our Binomial distribution with $p = 0.2$, the probability of getting infected:

$$\mathbb{P}(I = k) = \binom{6}{k}(0.2)^k(0.8)^{6-k}$$

We are asked to find the probability that only two patients contract the virus after having the received the vaccine:

$$\mathbb{P}(I = 2) = \binom{6}{2}(0.2)^2(0.8)^4 \approx 0.245$$

So, there is a 24.5% chance that 2 people will contract COVID-19 after having received the Pfizer vaccine. The plot of the distribution is shown in the following. As we can see, since the vaccine has a pretty high success rate (and therefore a lower failure rate, i.e. someone gets infected after having received the vaccine), the highest probabilities are associated with 0 or 1 patient actually becoming infected after receiving the vaccine, while the probability of more than 4 patients getting infected is very low.

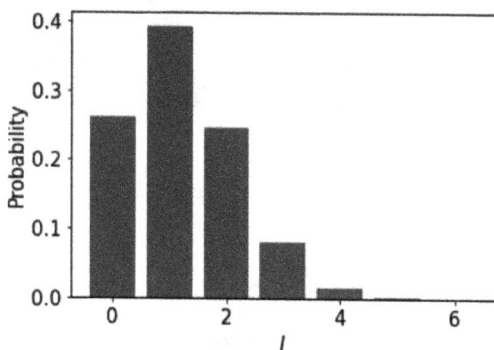

Calculating Expected Values

1. Looking back at our solution for exercise 1 in the previous section, we said that the random variable S can be 0, 1, 2 implying that we got no heads for both coin tosses, only one head or two heads both times. The probabilities associated with those values were

$$\mathbb{P}(0H) = \frac{1}{2} \cdot \frac{1}{2} = \frac{1}{4}$$

$$\mathbb{P}(1H) = 2 \cdot \frac{1}{2} \cdot \frac{1}{2} = \frac{1}{2}$$

$$\mathbb{P}(2H) = \frac{1}{2} \cdot \frac{1}{2} = \frac{1}{4}$$

So, the expected value would be

$$\mathbb{E}S = 0 \cdot \frac{1}{4} + 1 \cdot \frac{1}{2} + 2 \cdot \frac{1}{4} = 1$$

So, we expect to get 1 head when we toss a coin twice, which makes sense!

2. (a) The plot of the distribution is shown above.

   (b) The expected value can be computed as

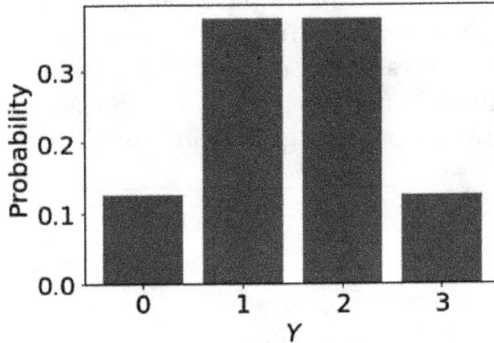

$$\mathbb{E}Y = 0 \cdot \frac{1}{8} + 1 \cdot \frac{3}{8} + 2 \cdot \frac{3}{8} + 3 \cdot \frac{1}{8} = 1.5$$

The distribution is symmetric, so the expected value which can also be viewed as the center of mass of a distribution should be at 1.5 in this case, which is what we have calculated.

Chapter 2

Eucledian Vectors

1. (a) In this case, our coordinate system consists of a plane where the axes are the $\hat{N}$ and $\hat{E}$ directions which are analogous to $\hat{y}$ and $\hat{x}$ and in and out of place direction is in this $\hat{F}$ direction. So, if we can write a displacement vector, $\vec{D}$, in terms of these axes, then

$$\vec{D} = 4\hat{E} + 5\hat{N} - 3\hat{F}$$

The magnitude of the vector is

$$|\vec{D}| = \sqrt{4^2 + 5^2 + (-3)^2} = \sqrt{50} \approx 7.1$$

(b) If we now have these new axes $\hat{NE}$ and $\hat{NW}$ that are 45° from the original axes, then we can see that these vectors are made of equal components of $\hat{E}$ and $\hat{N}$, so first we can write these new unit vectors in terms of the old ones. Normalizing the vectors, we find that

$$\hat{NE} = \frac{1}{\sqrt{2}}(\hat{N} + \hat{E})$$

$$\hat{NW} = \frac{1}{\sqrt{2}}(\hat{N} - \hat{E})$$

Then, we can re-express the original unit vectors in terms of the new unit vectors as

$$\hat{E} = \frac{1}{\sqrt{2}}(\hat{NE} - \hat{NW})$$

$$\hat{N} = \frac{1}{\sqrt{2}}(\hat{NE} + \hat{NW})$$

So, now, the movement $4\hat{E} = \frac{4}{\sqrt{2}}(\hat{NE} - \hat{NW})$ and $5\hat{E} = \frac{5}{\sqrt{2}}(\hat{NE} + \hat{NW})$. Thus, the total amount of movement in the $\hat{NE}$ and $\hat{NW}$ directions are $\frac{4}{\sqrt{2}} + \frac{5}{\sqrt{2}} = \frac{9}{\sqrt{2}}$ and $-\frac{4}{\sqrt{2}} + \frac{5}{\sqrt{2}} = \frac{1}{\sqrt{2}}$, respectively. Alternatively, you can take the dot product to figure out the *projection* of $4\hat{E}$ and $5\hat{N}$ in the new axes in the following way:

$$4\hat{E} \cdot \hat{NW} = 4\hat{E} \cdot \frac{1}{\sqrt{2}}(\hat{N} + \hat{E}) = \frac{4}{\sqrt{2}}$$

Noting here that $\hat{E} \cdot \hat{N} = 0$ since the vectors are orthogonal! Either way, the result will be the same. Movement in $\hat{F}$ axes remains $-3\hat{F}$. So, the vector written with these new axes is

$$\vec{D}' = \frac{9}{\sqrt{2}}\hat{NE} + \frac{1}{\sqrt{2}}\hat{NW} - 3\hat{F}$$

If we compute the magnitude of this vector,

$$|\vec{D}'| = \sqrt{\left(\frac{9}{\sqrt{2}}\right)^2 + \left(\frac{1}{\sqrt{2}}\right)^2 + (-3)^2} = \sqrt{50} \approx 7.1$$

This is an important conclusion in linear algebra, the magnitude of a vector does not depend on how you choose to write the vector. The unit vectors that we choose to write the vector form what is known as **basis** and here we can see that the magnitude or length of a vector is *basis independent*!

2. Using the rules of adding vectors, we simply add the components that are the same to each other, so

$$\vec{R} = \vec{P} + \vec{Q} = (1+6)\hat{i} + (3+3)\hat{j} + (4+2)\hat{k} = 7\hat{i} + 6\hat{j} + 6\hat{k}$$

Taking the magnitude of $\hat{R}$,

$$|\hat{R}| = \sqrt{7^2 + 6^2 + 6^2} = \sqrt{151} \approx 12.3$$

3. The point of this problem is to normalize vectors and use the dot product.

   (a) To normalize $\vec{u}$ and $\vec{v}$, we just need to divide the vectors by their magnitude, respectively:

   $$|\vec{u}| = \sqrt{2^2 + 3^2 + 4^2} = \sqrt{29}$$

   $$|\vec{v}| = \sqrt{2^2 + (-6)^2 + 7^2} = \sqrt{89}$$

   Taking the dot product, we have

   $$\frac{\vec{u}}{\sqrt{29}} \cdot \frac{\vec{v}}{\sqrt{89}} = \frac{2 \cdot 2 + 3 \cdot -6 + 4 \cdot 7}{\sqrt{29}\sqrt{89}} = \frac{14}{\sqrt{2581}} \approx 0.28$$

(b) We already know that $\hat{i} = \begin{pmatrix} 1 & 0 & 0 \end{pmatrix}$ and we just need to normalize $\vec{p}$ which has a magnitude of $\sqrt{17}$. However, we did not even need to do that here because we see that $\vec{p}$ has no component in the $\hat{i}$ direction at all, so their dot product is 0. Geometrically, $\vec{p}$ is located purely in the yz-plane.

Dirac Notation

1. The *bra* vector is the "dual" of the *ket* vector, i.e. the conjugate transpose of a *ket* vector:

$$\langle u| = ((|u\rangle)^*)^\top = \begin{pmatrix} (1-i) & 4 \end{pmatrix}$$

The * symbol means taking the conjugate of each element in the vector and the $^\top$ symbol means transposing or essentially making a column vector into a row vector. From a dimensions perspective, a column vector is 2×1 meaning 2 rows and 1 column. Transposing flips these dimensions, leading to 1×2, implying 1 row and 2 columns, which is a row vector!

2. The *ket* vector is the dual of the *bra* vector. Using the same idea as in (1),

$$\langle v| = \begin{pmatrix} 0.5+i \\ 0.2 \\ 1-0.5i \\ 0.2 \end{pmatrix}$$

3. This problem is very similar to what we did in the previous exercises with vectors in real vector spaces. We need to normalize and compute the inner product which is a generalized version of the dot product:

$$||P\rangle| = \sqrt{\langle P|P\rangle} = \sqrt{\begin{pmatrix} 1 & (2-i) & -i \end{pmatrix} \begin{pmatrix} 1 \\ 2+i \\ i \end{pmatrix}}$$

$$= \sqrt{1^2 + (2-i)(2+i) + (-i)(i)} = \sqrt{1 + (2^2 + 1^2) + 1} = \sqrt{7}$$

Similarly,

$$||P\rangle| = \sqrt{\langle P|P\rangle} = \sqrt{\frac{1}{2^2}(2^2 + 2^2) + 4^2 + (6^2 + 2^2))} = 4$$

So, the inner product of the normalized vectors is

$$\langle P|Q \rangle = \frac{1}{\sqrt{7}} \begin{pmatrix} 1 & (2-i) & -i \end{pmatrix} \frac{1}{2 \cdot 4} \begin{pmatrix} 2+2i \\ 4 \\ 6+2i \end{pmatrix}$$

$$= \frac{1}{8\sqrt{7}} \left((2+2i) + 4(2-i) - i(6+2i) \right) = \frac{12-8i}{8\sqrt{7}} \approx 0.57 - 0.38i$$

Superposition Principle

1. In this problem, the key is to determine whether the two vectors are scalar multiples of each other.

   (a) It is clear that $|v_1\rangle$ and $|v_2\rangle$ are linearly dependent since $3|v_1\rangle = |v_2\rangle$.

   (b) It is clear that there is no scalar that we multiply either vector to obtain the other.

2. In Q1, we found that the set of vectors from part b were linearly independent, and now, we can use these vectors to construct the vector $|\psi\rangle$. In this way, $|\psi\rangle$ is a linear combination of the two vectors, $|v_1\rangle$ and $|v_2\rangle$, where our job is to find what scaling factors achieve this:

$$|\psi\rangle = c_1 |v_1\rangle + c_2 |v_2\rangle$$

The more systematic way to do this is to actually use matrices, where the columns of the matrix are the linearly independent vectors. All that basically gets us is the following system of equations:

$$5c_1 + 10c_2 = 15$$

$$2c_1 + 2c_2 = 2$$

However, we can also figure this out through trial and error. It is easy to check that $c_1 = -1$ and $c_2 = 2$:

$$-\begin{pmatrix} 5 \\ 2 \end{pmatrix} + 2\begin{pmatrix} 10 \\ 2 \end{pmatrix} = \begin{pmatrix} 15 \\ 2 \end{pmatrix}$$

One thing the system of equations tells us is that there is only one unique solution here to express $|\psi\rangle$ in terms of these chosen linearly independent vectors.

3. We have learned now that a quantum state can be represented as a *complex* vector, which means that c_1 and c_2 can be complex numbers, thus there are actually 2 unknowns for each number: its real and imaginary parts. Now, we have a system of equations:

$$c_1 + c_2 = i$$

$$c_1 - c_2 = 1$$

Solving the system, we obtain that $c_1 = \frac{i+1}{2}$ and $c_2 = \frac{i-1}{2}$. To normalize, we just need the magnitude of $|\psi\rangle$:

$$\sqrt{\langle\psi|\psi\rangle} = \sqrt{|c_1|^2 + |c_2|^2} = \sqrt{2}$$

Thus, the normalized $|\psi\rangle$, labeled $|\hat{\psi}\rangle$, can be written as the following linear combination of the basis states:

$$|\hat{\psi}\rangle = \frac{1}{\sqrt{2}}|\psi\rangle = \frac{1}{\sqrt{2}}\left(\frac{i+1}{2}\begin{pmatrix}1\\1\end{pmatrix} + \frac{i-1}{2}\begin{pmatrix}1\\-1\end{pmatrix}\right)$$

4. We defined in the notes that if $|u\rangle$ and $|v\rangle$ are two vectors that have n elements, then their inner product is

$$\langle u|v\rangle = \begin{pmatrix} u_1^* & u_2^* & u_3^* & \cdots & u_n^* \end{pmatrix} \begin{pmatrix} v_1 \\ v_2 \\ v_3 \\ \vdots \\ v_n \end{pmatrix} = u_1^* v_1 + u_2^* v_2 + \cdots + u_n^* v_n$$

Orthogonality means that the inner product of the two vectors must be 0 if they are orthogonal and 1 if they are the same vector, i.e. $|u\rangle = |v\rangle$, since we are saying that both states are normalized. Then, for two indices i and j which just mean the ith element of $|u\rangle$ and jth element of $|v\rangle$, their inner product is

$$\langle u|v\rangle = \delta_{ij}$$

Matrices

1. Recall that tranposing a matrix means to flip it over its diagonal. So, we have that

$$\mathbf{A}^\mathsf{T} + \mathbf{B} + \mathbf{C} = \begin{pmatrix} -8 & 7 \\ -6 & 3 \end{pmatrix} + \begin{pmatrix} 9 & -1 \\ 5 & 0 \end{pmatrix} + \begin{pmatrix} 2 & -2 \\ 4 & 1 \end{pmatrix} = \begin{pmatrix} 3 & 4 \\ 3 & 4 \end{pmatrix}$$

2.

$$\mathbf{B} + \mathbf{D} = \begin{pmatrix} 9 & -1 \\ 5 & 0 \end{pmatrix} + \begin{pmatrix} 2 & -2 \\ 4 & 1 \\ 3 & 0 \end{pmatrix}$$

This computation cannot be carried out because the matrices do not have the same dimensions!

3.

$$\mathbf{A}\mathbf{D}^{\mathsf{T}} = \begin{pmatrix} -8 & -6 \\ 7 & 3 \end{pmatrix} \begin{pmatrix} 2 & 4 & 3 \\ -2 & 1 & 0 \end{pmatrix}$$

$$= \begin{pmatrix} -8(2) - 6(-2) & -8(4) - 6(1) & -8(3) - 6(0) \\ 7(2) + 3(-2) & 7(4) + 3(1) & 7(3) + 3(0) \end{pmatrix}$$

$$= \begin{pmatrix} -4 & -38 & -24 \\ 8 & 31 & 21 \end{pmatrix}$$

4.

$$\mathbf{D}\mathbf{A} = \begin{pmatrix} 2 & -2 \\ 4 & 1 \\ 3 & 0 \end{pmatrix} \begin{pmatrix} -8 & -6 \\ 7 & 3 \end{pmatrix}$$

$$= \begin{pmatrix} 2(-8) - 2(7) & 2(-6) - 2(3) \\ 4(-8) + 1(7) & 4(-6) + 1(3) \\ 3(-8) + 0(7) & 3(-6) + 0(3) \end{pmatrix}$$

$$= \begin{pmatrix} -30 & -18 \\ -25 & -21 \\ -24 & -18 \end{pmatrix}$$

5.

$$3\mathbf{C} = 3 \begin{pmatrix} 2 & -2 \\ 4 & 1 \end{pmatrix} = \begin{pmatrix} 6 & -6 \\ 12 & 3 \end{pmatrix}$$

Linear Transformations

1. The matrix $\mathbf{X}$ acting on $|u\rangle$ results in the following:

$$\frac{1}{\sqrt{2}}\begin{pmatrix} 0 & 1 \\ 1 & 0 \end{pmatrix}\begin{pmatrix} 1 \\ i \end{pmatrix} = \frac{1}{\sqrt{2}}\begin{pmatrix} i \\ 1 \end{pmatrix}$$

Let's try to factor out an i to see if we can get $|u\rangle$ back just scaled by some number. Recall that $1/i = -i$, so we have that

$$\mathbf{X}|u\rangle = \frac{i}{\sqrt{2}}\begin{pmatrix} 1 \\ -i \end{pmatrix}$$

Thus, $|u\rangle$ is not an eigenvector of $\mathbf{X}$.

2. The matrix $\mathbf{Y}$ acting on $|u\rangle$ results in the following:

$$\frac{1}{\sqrt{2}}\begin{pmatrix} 0 & -i \\ i & 0 \end{pmatrix}\begin{pmatrix} 1 \\ i \end{pmatrix} = \frac{1}{\sqrt{2}}\begin{pmatrix} 1 \\ i \end{pmatrix} = |u\rangle$$

Thus, $|u\rangle$ *is* an eigenvector of $\mathbf{Y}$ with eigenvalue $+1$.

3. The matrix $\mathbf{Z}$ acting on $|\psi\rangle$ results in the following:

$$\frac{1}{\sqrt{2}}\begin{pmatrix} 1 & 0 \\ 0 & -1 \end{pmatrix}\begin{pmatrix} c_1 \\ c_2 \end{pmatrix} = \begin{pmatrix} c_1 \\ -c_2 \end{pmatrix} \neq |\psi\rangle$$

Thus, $|\psi\rangle$ is not an eigenvector of $\mathbf{Z}$. However, $|0\rangle = \begin{pmatrix} 1 \\ 0 \end{pmatrix}$ and $|1\rangle = \begin{pmatrix} 0 \\ 1 \end{pmatrix}$ states are.

Calculating Eigenvalues and Eigenvectors

1. The eigenvalues for matrix $\mathbf{D}$ were found in the corresponding section to be $\lambda = \pm 1$. In order to find eigenvectors, we need to find the vectors that satisfy the following:

$$\mathbf{D}|e\rangle = \lambda|u\rangle$$

$$\begin{pmatrix} 1 & 0 \\ 1 & -1 \end{pmatrix}\begin{pmatrix} a \\ b \end{pmatrix} = \pm\begin{pmatrix} a \\ b \end{pmatrix}$$

where a and b are the unknown elements of the eigenvector, which can be found by using one of the linear equations. In this case, we take the equation from the second row of the matrix:

$$a - b = \pm b$$

Thus, $a = 2b$ or $a = 0$. In the first case, for any choice of b, a is twice that. For the second case, a is always 0, so b can take on any value. So the two eigenvectors are

$$|u_1\rangle = \begin{pmatrix} 2 \\ 1 \end{pmatrix} \quad |u_2\rangle = \begin{pmatrix} 0 \\ 1 \end{pmatrix}$$

Let's prove that they give us what we want:

$$\begin{pmatrix} 1 & 0 \\ 1 & -1 \end{pmatrix}\begin{pmatrix} 2 \\ 1 \end{pmatrix} = \begin{pmatrix} 2 \\ 1 \end{pmatrix}$$

So, $|u_1\rangle$ is the eigenvector associated with the eigenvalue $+1$:

$$\begin{pmatrix} 1 & 0 \\ 1 & -1 \end{pmatrix}\begin{pmatrix} 0 \\ 1 \end{pmatrix} = \begin{pmatrix} 0 \\ -1 \end{pmatrix} = -\begin{pmatrix} 0 \\ 1 \end{pmatrix}$$

So, $|u_2\rangle$ is the eigenvector associated with the eigenvalue -1.

2. Now, we follow the same procedure to find the eigenvalues and eigenvectors of matrix $\mathbf{P}$. Once again, we want to solve the following eigenvalue equation:

$$\mathbf{P}|u\rangle = \lambda|u\rangle$$

First, let's compute the determinant of this matrix to find the characteristic equation:

$$\det(\mathbf{P} - \lambda\mathbb{I}) = \begin{vmatrix} 1-\lambda & 1 \\ 1 & -1-\lambda \end{vmatrix} = (1-\lambda)(-1-\lambda) - 1 = 0$$

Expanding out the polynomial, we get

$$\lambda^2 - 2 = 0$$

$$\lambda = \pm\sqrt{2}$$

Now, we can find the eigenvectors by plugging back the eigenvalues into the eigenvalue problem:

$$\begin{pmatrix} 1 & 1 \\ 1 & -1 \end{pmatrix}\begin{pmatrix} a \\ b \end{pmatrix} = \pm\sqrt{2}\begin{pmatrix} a \\ b \end{pmatrix}$$

Taking the equation formed by the first row,

$$a + b = \pm\sqrt{2}a$$

So, we get that $b = (\sqrt{2} - 1)a$ or $b = -(\sqrt{2} + 1)a$. If we take $a = 1$, then we obtain the following eigenvectors:

$$|u_1\rangle = \begin{pmatrix} 1 \\ (\sqrt{2} - 1) \end{pmatrix} \qquad |u_2\rangle = \begin{pmatrix} 1 \\ -(\sqrt{2} + 1) \end{pmatrix}$$

Let's test that we obtain the expected result:

$$\mathbf{P}|u_1\rangle = \begin{pmatrix} 1 & 1 \\ 1 & -1 \end{pmatrix}\begin{pmatrix} 1 \\ (\sqrt{2} - 1) \end{pmatrix} = \begin{pmatrix} \sqrt{2} \\ (2 - \sqrt{2}) \end{pmatrix} = \sqrt{2}\begin{pmatrix} 1 \\ (\sqrt{2} - 1) \end{pmatrix}$$

$$\mathbf{P}|u_2\rangle = \begin{pmatrix} 1 & 1 \\ 1 & -1 \end{pmatrix}\begin{pmatrix} 1 \\ -(\sqrt{2} + 1) \end{pmatrix} = \begin{pmatrix} -\sqrt{2} \\ (2 + \sqrt{2}) \end{pmatrix} = -\sqrt{2}\begin{pmatrix} 1 \\ -(\sqrt{2} + 1) \end{pmatrix}$$

Properties of Unitary Matrices

1. First, to check if the matrix is unitary, we must verify that $\mathbf{Y}^\mathsf{T}\mathbf{Y} = \mathbb{I}$.

$$\begin{bmatrix} 0 & -i \\ i & 0 \end{bmatrix}\begin{bmatrix} 0 & -i \\ i & 0 \end{bmatrix} = \begin{bmatrix} 1 & 0 \\ 0 & 1 \end{bmatrix} = \mathbb{I}$$

We have shown that the matrix is unitary, now let us use the characteristic equation to find the eigenvalues:

$$\det(\mathbf{Y} - \lambda\mathbb{I}) = 0 = \begin{vmatrix} -\lambda & -i \\ i & -\lambda \end{vmatrix} = -\lambda^2 - 1 = 0$$

So, the eigenvalues are clearly $\lambda = \pm 1$. Now, we can solve the eigenvectors by plugging the eigenvalues back into the eigenvalue problem:

$$\mathbf{Y}|u\rangle = \pm|u\rangle$$

$$\begin{pmatrix} 0 & -i \\ i & 0 \end{pmatrix}\begin{pmatrix} a \\ b \end{pmatrix} = \pm\begin{pmatrix} a \\ b \end{pmatrix}$$

Taking the equation from the second row, we have

$$ia = \pm b$$

which means if $b = \pm 1$, then $a = i$. So, our two eigenvectors are

$$|u_1\rangle = \begin{pmatrix} i \\ 1 \end{pmatrix} \qquad |u_2\rangle = \begin{pmatrix} i \\ -1 \end{pmatrix}$$

2. First, let's check if **F** is unitary:

$$\begin{bmatrix} 0 & -i \\ 1 & 0 \end{bmatrix} \begin{bmatrix} 0 & 1 \\ i & 0 \end{bmatrix} = \begin{bmatrix} 1 & 0 \\ 0 & 1 \end{bmatrix} = \mathbb{I}$$

We have shown that the matrix is unitary, now let us use the characteristic equation to find the following eigenvalues:

$$\det(\mathbf{F} - \lambda\mathbb{I}) = 0 = \begin{vmatrix} -\lambda & i \\ 1 & -\lambda \end{vmatrix} = -\lambda^2 + i = 0$$

So, the eigenvalues are clearly $\lambda = \pm\sqrt{i}$. Recalling complex exponentials, we know that $e^{i\frac{\pi}{2}} = \cos\frac{\pi}{2} + i\sin\frac{\pi}{2} = i$. Therefore, $\sqrt{i} = e^{i\frac{\pi}{4}} = \frac{1+i}{\sqrt{2}}$. Now, we can solve the eigenvectors by plugging the eigenvalues back into the eigenvalue problem:

$$\mathbf{F}|u\rangle = \pm|u\rangle$$

$$\begin{pmatrix} 0 & 1 \\ i & 0 \end{pmatrix}\begin{pmatrix} a \\ b \end{pmatrix} = \pm\left(\frac{1+i}{\sqrt{2}}\right)\begin{pmatrix} a \\ b \end{pmatrix}$$

Taking the equation from the first row, we have

$$b = \pm\left(\frac{1+i}{\sqrt{2}}\right)a$$

which means if $a = 1$, then $b = \pm\left(\frac{1+i}{\sqrt{2}}\right)$. So, our two eigenvectors are

$$|u_1\rangle = \begin{pmatrix} 1 \\ \left(\frac{1+i}{\sqrt{2}}\right) \end{pmatrix} \qquad |u_2\rangle = \begin{pmatrix} 1 \\ -\left(\frac{1+i}{\sqrt{2}}\right) \end{pmatrix}$$

Chapter 3

Introduction to Quantum Mechanics

1. (a) Based on the definition of the Schrodinger cat paradox, the independent states for the radioactive substance are "decayed," $|1\rangle$ or "not decayed" $|0\rangle$.

(b) We know that each of the independent states listed in part a have a 50% chance of occurring. We also understand that we can express the general state of the quantum system in terms of the basis states, which in this case are the independent vectors $|0\rangle$ and $|1\rangle$. We can start with the following guess for the general superposition state:

$$|\psi\rangle = \frac{1}{2}|0\rangle + \frac{1}{2}|1\rangle$$

However, this state is not normalized since $\langle\psi|\psi\rangle \neq 1$. So, we use the normalization condition to solve for the normalization factor, c:

$$|\psi\rangle = \frac{c}{2}(|0\rangle + |1\rangle)$$

$$\langle\psi|\psi\rangle = \left|\frac{c}{2}\right|^2 + \left|\frac{c}{2}\right|^2 = 1$$

Therefore, we can solve that c is

$$2\left|\frac{c}{2}\right|^2 = 1$$

$$c = \frac{2}{\sqrt{2}}$$

Plugging this in, we have that $|\psi\rangle$ is

$$|\psi\rangle = \frac{1}{\sqrt{2}}|0\rangle + \frac{1}{\sqrt{2}}|1\rangle$$

2. The main point that Schrodinger was trying to express with this proposed paradox is that the "conscious observer"-driven interpretation of the collapse of the superposition state was wrong, and it was shown to be wrong experimentally as well. So, if we were to consider the radioactive substance in the situation described, the superposition would collapse due to the interaction of the radioactive substance with the Geiger count, the measurement device. The superposition would not collapse to a definite state just because a conscious observer opened the steel chamber. Again, the lesson here is that quantum superposition is disturbed by the interaction of the quantum state with ANYTHING in the environment.

Spins in Magnetic Fields

1. If the source of silver atoms were sent through an SG apparatus whose field is mostly oriented in the y-direction, we would still observe atoms deflect either up or down leading to two distinct spots on the screen, indicating $|\uparrow\rangle$ or $|\downarrow\rangle$.

2. In this setup, there is a beam of silver atoms which have random spin orientations and the beam goes through the first SG apparatus, oriented in the z-direction. The spins are measured to be one of the basis states $|\uparrow\rangle$ or $|\downarrow\rangle$, indicating "spin-up" or "spin-down," respectively. Then, the "spin-up" pile is redirected through another SG apparatus where the field is oriented in the z-direction. The resulting measurement will only yield spin-up atoms with 100% probability, since the states sent to the second SG apparatus were definitely $|\uparrow\rangle$ which is a basis state of the SG apparatus whose field is oriented in the z-direction.

3. Once again, we have a a beam of silver atoms which have random spin orientations and the beam goes through the first SG apparatus, oriented in the z-direction. The spins are measured to be one of the basis states $|\uparrow\rangle$ or $|\downarrow\rangle$ indicating "spin-up" or "spin-down," respectively. Then, the "spin-up" pile is redirected through an SG apparatus whose field is oriented in the x-direction. The basis states that the apparatus can measure are $|\rightarrow\rangle$ and $|\leftarrow\rangle$. We know that if we are sending in only the "spin-up" pile, there is a 50% chance of getting either $|\rightarrow\rangle$ and $|\leftarrow\rangle$. Finally, we redirect just the $|\rightarrow\rangle$ pile to another SG apparatus, whose field is oriented in the z-direction. Now, the basis states are $|\uparrow\rangle$ or $|\downarrow\rangle$, so sending in $|\rightarrow\rangle$, we will get either $|\uparrow\rangle$ or $|\downarrow\rangle$ with 50% probability for each case. So, the final output is a superposition of $|\uparrow\rangle$ or $|\downarrow\rangle$ each with 50% probability of being measured. Considering the entire sequence from the beginning, we have the original beam where 50% of it ends up being $|\uparrow\rangle$, and then redirected through the second SG apparatus, there is 25% chance for the original beam to be $|\rightarrow\rangle$, and finally, there is a 12.5% chance to measure $|\uparrow\rangle$ or $|\downarrow\rangle$.

Light Polarization

1. From the notes, we know that the vertical polarization state is $|\uparrow\rangle = \begin{pmatrix} 0 \\ 1 \end{pmatrix}$, and we also know that the vectors $|\nearrow\rangle$ and $|\searrow\rangle$ can be

written in the $\{|\rightarrow\rangle, |\uparrow\rangle\}$ basis as follows:

$$|\nearrow\rangle = |D\rangle = \frac{1}{\sqrt{2}}(|\rightarrow\rangle + |\uparrow\rangle) = \frac{1}{\sqrt{2}}\begin{pmatrix} 1 \\ 1 \end{pmatrix}$$

$$|\searrow\rangle = |A\rangle = \frac{1}{\sqrt{2}}(|\rightarrow\rangle - |\uparrow\rangle) = \frac{1}{\sqrt{2}}\begin{pmatrix} 1 \\ -1 \end{pmatrix}$$

If we take $|\nearrow\rangle - |\searrow\rangle$, we end up with

$$|\nearrow\rangle - |\searrow\rangle = \frac{2}{\sqrt{2}}|\uparrow\rangle$$

Solving for $|\uparrow\rangle$ and noting that $\frac{\sqrt{2}}{2} = \frac{1}{\sqrt{2}}$, we get that

$$|\uparrow\rangle = \frac{1}{\sqrt{2}}(|\nearrow\rangle - |\searrow\rangle)$$

We could also have seen this visually.

2. Again, from the notes, we know that the horizontal polarization state is $|\rightarrow\rangle = \begin{pmatrix} 1 \\ 0 \end{pmatrix}$, and we also know that the vectors $|\circlearrowright\rangle$ and $|\circlearrowleft\rangle$ can be written in the $\{|\rightarrow\rangle, |\uparrow\rangle\}$ basis as follows:

$$|\circlearrowright\rangle = |R\rangle = \frac{1}{\sqrt{2}}(|\rightarrow\rangle - i|\uparrow\rangle) = \frac{1}{\sqrt{2}}\begin{pmatrix} 1 \\ -i \end{pmatrix}$$

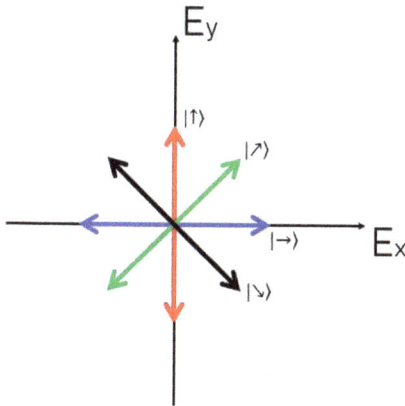

$$|\circlearrowright\rangle = |L\rangle = \frac{1}{\sqrt{2}}(|\rightarrow\rangle + i\,|\uparrow\rangle) = \frac{1}{\sqrt{2}}\begin{pmatrix} 1 \\ i \end{pmatrix}$$

By observation, we see that taking a sum of $|\circlearrowleft\rangle + |\circlearrowright\rangle$, we have

$$|\circlearrowleft\rangle + |\circlearrowright\rangle = \frac{2}{\sqrt{2}}\,|\rightarrow\rangle$$

Solving for $|\rightarrow\rangle$ and noting that $\frac{\sqrt{2}}{2} = \frac{1}{\sqrt{2}}$, we get

$$|\rightarrow\rangle = \frac{1}{\sqrt{2}}\,(|\circlearrowleft\rangle + |\circlearrowright\rangle)$$

Measurement with Light

The state emitted from the laser is initially $|\psi\rangle = |\nearrow\rangle$, so 50% goes through the $|\uparrow\rangle$ polarizer and now is projected into the $|\uparrow\rangle$ state. After passing through the $|\nearrow\rangle$ polarizer, the state becomes

$$|\psi'\rangle = \frac{1}{\sqrt{2}}(|\rightarrow\rangle + |\uparrow\rangle)$$

So, when this state goes through the $|\rightarrow\rangle$ polarizer, $|\psi\rangle$ will be projected onto $|\rightarrow\rangle$:

$$\langle\psi|\rightarrow\rangle = \frac{1}{\sqrt{2}}(\langle\rightarrow|\rightarrow\rangle + \langle\uparrow|\rightarrow\rangle)$$

But we know that $|\uparrow\rangle$ and $|\rightarrow\rangle$ are orthogonal states, so $\langle\uparrow|\rightarrow\rangle) = 0$ and $\langle\rightarrow|\rightarrow\rangle = 1$. Thus,

$$\langle\psi|\rightarrow\rangle = \frac{1}{\sqrt{2}}$$

This square of this result gives us the probability, which is $\frac{1}{2}$. This means that only 50% of the state $|\psi'\rangle$ actually makes it through the $|\rightarrow\rangle$ polarizer. So, considering that we started with $|\psi\rangle = |\nearrow\rangle$, which went through the $|\uparrow\rangle$ polarizer by 50%, it was then only 50%

that made it through the $|\nearrow\rangle$, and finally, only 50% of that made it through the $|\rightarrow\rangle$ polarizer. Thus, only 12.5% of the original state can actually be observed after passing through all the polarizers.

Chapter 4

Controlling Qubits

1. The Pauli X, Y, and Z matrices are

$$\mathbf{X} = \begin{pmatrix} 0 & 1 \\ 1 & 0 \end{pmatrix} \quad \mathbf{Y} = \begin{pmatrix} 0 & -i \\ i & 0 \end{pmatrix} \quad \mathbf{Z} = \begin{pmatrix} 1 & 0 \\ 0 & -1 \end{pmatrix} \quad \mathbf{I} = \begin{pmatrix} 1 & 0 \\ 0 & 1 \end{pmatrix}$$

We are just being asked to verify that the eigenvectors are as given in Section 4.2.1. Another way to label the Pauli matrices is using the symbol σ with a subscript $i = X, Y, Z$. So, to verify the eigenvectors, we know that they need to satisfy the eigenvalue equation:

$$\sigma_i |u\rangle = \lambda |u\rangle$$

For X, we have

$$\begin{pmatrix} 0 & 1 \\ 1 & 0 \end{pmatrix} \begin{pmatrix} \frac{1}{\sqrt{2}} \\ \pm\frac{1}{\sqrt{2}} \end{pmatrix} = \pm 1 \cdot \begin{pmatrix} \frac{1}{\sqrt{2}} \\ \pm\frac{1}{\sqrt{2}} \end{pmatrix}$$

$$\begin{pmatrix} \pm\frac{1}{\sqrt{2}} \\ \frac{1}{\sqrt{2}} \end{pmatrix} = \pm 1 \cdot \begin{pmatrix} \frac{1}{\sqrt{2}} \\ \pm\frac{1}{\sqrt{2}} \end{pmatrix}$$

Distributing the ± 1, we see that both sides of the equation are equal. Similarly, for Y, we have

$$\begin{pmatrix} 0 & -i \\ i & 0 \end{pmatrix} \begin{pmatrix} \frac{1}{\sqrt{2}} \\ \pm\frac{i}{\sqrt{2}} \end{pmatrix} = \pm 1 \cdot \begin{pmatrix} \frac{1}{\sqrt{2}} \\ \pm\frac{i}{\sqrt{2}} \end{pmatrix}$$

Doing out the matrix multiplication and remembering that $-i \cdot i = 1$, we get

$$\begin{pmatrix} \pm\frac{1}{\sqrt{2}} \\ \frac{i}{\sqrt{2}} \end{pmatrix} = \pm 1 \cdot \begin{pmatrix} \frac{1}{\sqrt{2}} \\ \pm\frac{i}{\sqrt{2}} \end{pmatrix}$$

Distributing the ±1, we see that both sides of the equation are equal. The same procedure is followed to verify the eigenvectors of Z.

2. In Section 4.2.3, the reference point of deriving the general unitary transformation was that any single-qubit state in the $\{|0\rangle, |1\rangle\}$ basis can be written as

$$|\psi\rangle = \cos\frac{\theta}{2}|0\rangle + \sin\frac{\theta}{2}e^{i\varphi}|1\rangle$$

which let us say that there is some unitary transformation that takes $|0\rangle = \binom{1}{0}$ to $|\psi\rangle$. So, if we write the transformation in the $\{|0\rangle, |1\rangle\}$ basis, then our starting point is

$$U = \begin{pmatrix} \cos(\theta/2) & a \\ e^{i\varphi}\sin(\theta/2) & b \end{pmatrix}$$

where a and b are complex numbers which we have to solve for. We use the fact that unitary matrices satisfy $U^{\dagger}U = I$. Writing this down, we have

$$U^{\dagger}U = \begin{pmatrix} \cos(\theta/2) & e^{-i\varphi}\sin(\theta/2) \\ a^* & b^* \end{pmatrix}\begin{pmatrix} \cos(\theta/2) & a \\ \sin(\theta/2)e^{i\varphi} & b \end{pmatrix} = \begin{pmatrix} 1 & 0 \\ 0 & 1 \end{pmatrix}$$

Now, let us do the matrix multiplication on the left-hand side:

$$\begin{pmatrix} \cos^2(\theta/2)\sin^2(\theta/2) & \cos(\theta/2)\cdot a + e^{-i\varphi}\sin(\theta/2)\cdot b \\ \cos(\theta/2)\cdot a^* + e^{-i\varphi}\sin(\theta/2)\cdot b^* & a^*a + b^*b \end{pmatrix}$$

$$= \begin{pmatrix} 1 & 0 \\ 0 & 1 \end{pmatrix}$$

The first matrix element $\cos^2(\theta/2)\sin^2(\theta/2) = 1$, so that doesn't tell us anything. The rest of the matrix elements give us the following three equations:

$$\cos(\theta/2)\cdot a + e^{-i\varphi}\sin(\theta/2)\cdot b = 0$$

$$\cos(\theta/2)\cdot a^* + e^{-i\varphi}\sin(\theta/2)\cdot b^* = 0$$

$$a^*a + b^*b = 1$$

Using the first equation, we can rearrange it to solve for b in terms of a:

$$b = -\cot(\theta/2)\, e^{i\varphi} \cdot a^{1}$$

The complex conjugate b^* is

$$b^* = -\cot(\theta/2)\, e^{-i\varphi} \cdot a^*$$

Plugging these two equations into the third equation $a^*a + b^*b = 1$, we get

$$a^*a\left(1 + \cot^2(\theta/2)\right) = 1$$

Using the trig identity $\csc^2\theta - \cot^2 = 1$, where $\csc\theta = \frac{1}{\sin\theta}$, we can simplify the equation to get

$$a^*a = |a|^2 = \sin^2(\theta/2)$$

Thus, the magnitude of $|a|$ is $\sin(\theta/2)$, however a itself can have some phase, so to be completely general, we must say that $a = \sin(\theta/2)\, e^{i\lambda}$, where λ is some arbitrary phase. Plugging this back into the equation for b, we get

$$b = -\cot(\theta/2)\, e^{i\varphi} \cdot \sin(\theta/2)\, e^{i\lambda}$$

which simplifies to

$$b = -\cos(\theta/2)\, e^{i(\varphi+\lambda)}$$

---

[1] Here, we have used the definition of $\cot\theta = \frac{\cos\theta}{\sin\theta}$ and the rule for exponentials $\frac{1}{e^{i\varphi}} = e^{-i\varphi}$ to simplify the equation.

Describing Measurements

1. The solution is shown in Fig. D.1.

```
q = QuantumRegister(1) #Initialize Quantum Register
c = ClassicalRegister(1) #Initialize Classical Register for measurement results
circ = QuantumCircuit(q,c) #Create Quantum Circuit
circ.ry(pi/4,0) #45 deg rotation about Y-axis
circ.rz(pi/4,0) #45 deg rotation about Z-axis

state = Statevector(circ)
print(state)
plot_bloch_multivector(state)
```

```
Statevector([0.85355339-0.35355339j, 0.35355339+0.14644661j],
            dims=(2,))
```

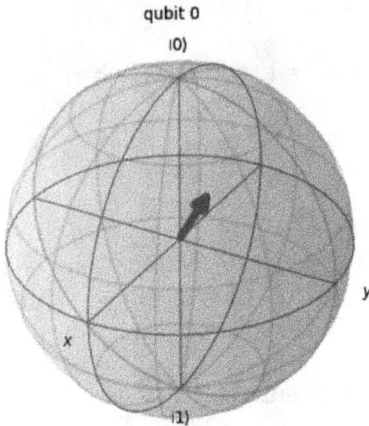

Figure D.1.   Solution to Question 1.

2. First, let's make copies of the circuit and do the relevant rotations to measure each component:

```
from copy import deepcopy

circX = deepcopy(circ)
circX.h(0)
circX.measure(q,c)

circY = deepcopy(circ)
circY.rx(pi/2,0)
circY.measure(q,c)

circZ = deepcopy(circ)
circZ.measure(q,c)
```

The results are shown in Figs. D.2–D.4. The resulting xyz vector is $\begin{pmatrix} 0.52 & 0.52 & 0.62 \end{pmatrix}$.

```
shots = 100
simulator = Aer.get_backend('qasm_simulator')
circ_transpile = transpile(circX, backend = simulator)
result = simulator.run(circ_transpile,shots = shots).result()
counts = result.get_counts()
print(counts)
plot_histogram(counts)
```

{'1': 24, '0': 76}

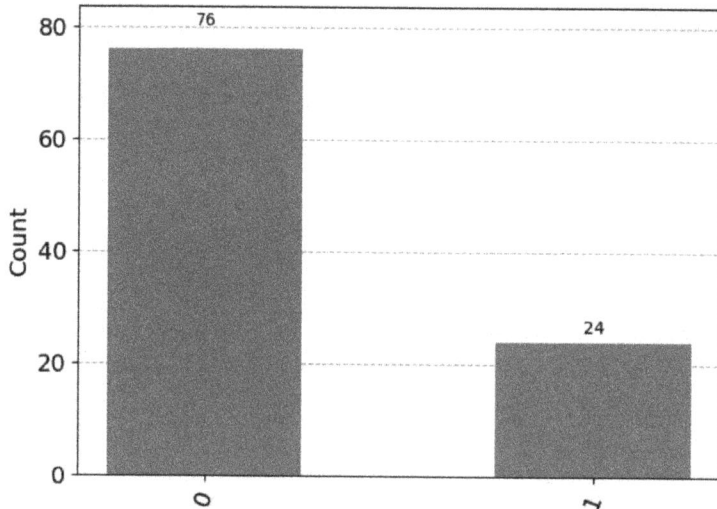

Figure D.2. Solution to Question 2 for X-component.

3. We can convert the x-, y-, z-components into angles $\theta = \arccos z$, and $\phi = \arctan \frac{y}{x}$. Plugging in, we get $\theta = 0.9$ rad or $\approx 52°$ and $\phi = \pi/4$ rad or $45°$. The Bloch vector is $\begin{pmatrix} \cos(0.9/2) \\ \sin(0.9/2)e^{i\pi/4} \end{pmatrix} \approx$ $\begin{pmatrix} 0.9 \\ 0.3 + 0.3i \end{pmatrix}$. This is the difference from the Bloch vector output by Statevector. In reality, there is an overall global phase of about $-\frac{\pi}{8}$ rad which resolves this seeming inconsistency. Remember that the global phase will not affect measurement outcomes!

```
shots = 100
simulator = Aer.get_backend('qasm_simulator')
circ_transpile = transpile(circY, backend = simulator)
result = simulator.run(circ_transpile,shots = shots).result()
counts = result.get_counts()
print(counts)
plot_histogram(counts)
```

{'1': 24, '0': 76}

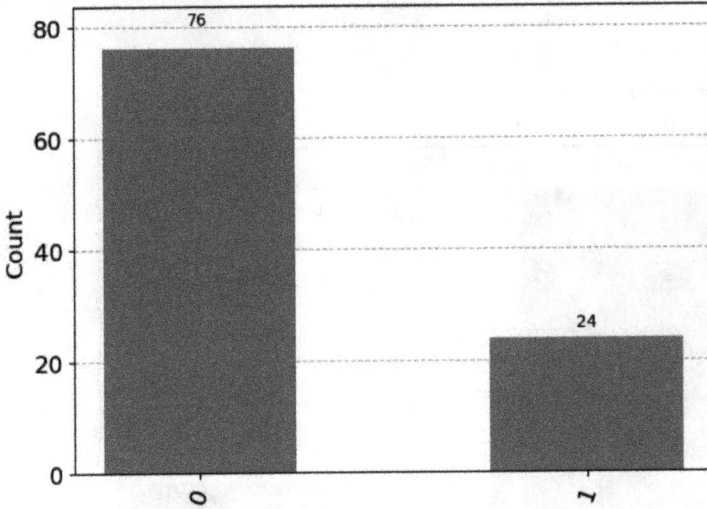

Figure D.3.   Solution to Question 2 for Y-component.

4. Computing the magnitude of the vector, we obtained

$$\sqrt{0.52^2 + 0.52^2 + 0.62^2} \approx 0.96$$

The value does not exactly equal one because we always have some statistical error since we cannot measure an infinite number of times.

5. Using 1000 shots, we get an updated xyz vector, (0.508 0.538 0.692), which has a magnitude

$$\sqrt{0.508^2 + 0.538^2 + 0.692^2} \approx 1$$

Thus, having more shots makes the vector more accurate!

```
shots = 100
simulator = Aer.get_backend('qasm_simulator')
circ_transpile = transpile(circZ, backend = simulator)
result = simulator.run(circ_transpile,shots = shots).result()
counts = result.get_counts()
print(counts)
plot_histogram(counts)
```

{'0': 81, '1': 19}

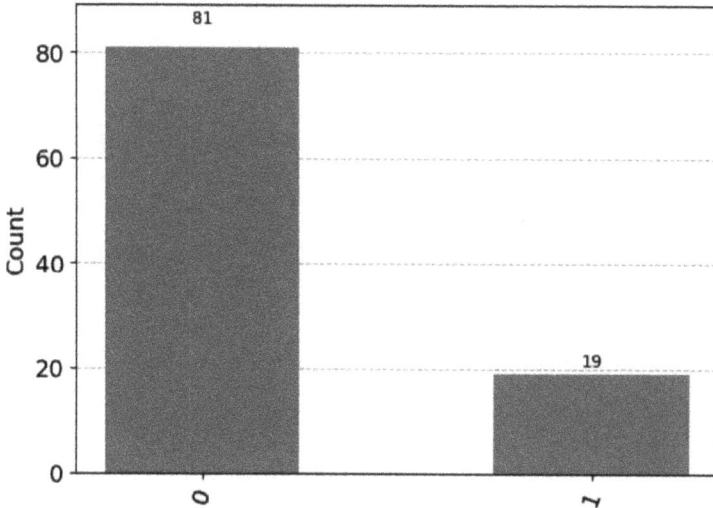

Figure D.4. Solution to Question 2 for Z-component.

Chapter 6

Product states

1. Using our knowledge of the tensor product, we can compute

$$|00\rangle = \begin{pmatrix} 1 \times \begin{pmatrix} 1 \\ 0 \end{pmatrix} \\ 0 \times \begin{pmatrix} 1 \\ 0 \end{pmatrix} \end{pmatrix} = \begin{pmatrix} 1 \\ 0 \\ 0 \\ 0 \end{pmatrix}$$

$$|01\rangle = \begin{pmatrix} 1 \times \begin{pmatrix} 0 \\ 1 \end{pmatrix} \\ 0 \times \begin{pmatrix} 0 \\ 1 \end{pmatrix} \end{pmatrix} = \begin{pmatrix} 0 \\ 1 \\ 0 \\ 0 \end{pmatrix}$$

$$|10\rangle = \begin{pmatrix} 0 \times \begin{pmatrix} 1 \\ 0 \end{pmatrix} \\ 1 \times \begin{pmatrix} 1 \\ 0 \end{pmatrix} \end{pmatrix} = \begin{pmatrix} 0 \\ 0 \\ 1 \\ 0 \end{pmatrix}$$

$$|11\rangle = \begin{pmatrix} 0 \times \begin{pmatrix} 0 \\ 1 \end{pmatrix} \\ 1 \times \begin{pmatrix} 0 \\ 1 \end{pmatrix} \end{pmatrix} = \begin{pmatrix} 0 \\ 0 \\ 0 \\ 1 \end{pmatrix}$$

2. Let's ensure that all the sum of the squared magnitude of all the coefficients is equal to 1. For the first state,

$$||\psi\rangle| = \sqrt{\frac{1}{2}^2 + \frac{1}{2}^2 + \frac{1}{2}^2 + \frac{1}{2}^2} = \sqrt{\frac{1}{4} + \frac{1}{4} + \frac{1}{4} + \frac{1}{4}} = 1$$

For the second state,

$$||\psi\rangle| = \sqrt{\frac{1}{2}^2 + \frac{1}{4}^2} = \sqrt{\frac{1}{4} + \frac{1}{16}} = \frac{\sqrt{5}}{4} \approx 0.56$$

So, this state is not normalized. To normalize it, we should divide by the magnitude we just computed. Then the normalized state becomes

$$|\psi\rangle_N = \frac{|\psi\rangle}{||\psi\rangle|} = \frac{1}{\sqrt{5}} (2|00\rangle + |01\rangle)$$

3. Using the property of the tensor product that we discussed, $(A \otimes B)(|u\rangle \otimes |v\rangle) = (A|u\rangle) \otimes (B|v\rangle)$, the first product is

$$X \otimes Z(|0\rangle \otimes |+\rangle) = X|0\rangle \otimes Z|+\rangle = |1\rangle \otimes |-\rangle$$

The second product is

$$I \otimes Y(|1\rangle \otimes |-\rangle) = I|1\rangle \otimes Y|-\rangle = |1\rangle \otimes i|+\rangle$$

We can compute out $Y|-\rangle$ since it is not obvious:

$$Y|-\rangle = \frac{1}{\sqrt{2}} \begin{pmatrix} 0 & -i \\ i & 0 \end{pmatrix} \begin{pmatrix} 1 \\ -1 \end{pmatrix} = \frac{1}{\sqrt{2}} \begin{pmatrix} i \\ i \end{pmatrix} = \frac{i}{\sqrt{2}} \begin{pmatrix} 1 \\ 1 \end{pmatrix} = i|+\rangle$$

Entangled states

1. In this problem, we will prepare all the Bell states using Qiskit and then visualize the states using the "plot state city" function. This function is a way to visualize the density matrix. Instead of using the Statevector to represent the state, we could also write the state in matrix form. It turns out that this is a more general way to write down any quantum state, not just the pure quantum states that we have working with in this book. A pure state means that you could write down a Statevector. However, in realistic situations, where noise is present, quatum states are not pure. For example, suppose we have a $|\Phi_+\rangle$ Bell state, but because of noise, there is some small percentage of the $|01\rangle$ state. We could not really write down a Statevector for this situation, however, we could write down a density matrix. For pure states, the density matrix, ρ, is simply the outer product[2]:

$$\rho = |\psi\rangle \langle\psi|$$

Let's start with the $|\Phi_+\rangle$ state. The circuit is given in Fig. D.5 and the plot from the "plot state city" function in Fig. D.6. As you can see, the matrix tells us that we have a probability of 0.5 to measure the state in $|00\rangle$ or $|11\rangle$.

---

[2] As a quick example for a single-qubit state, $|0\rangle$, the density matrix would be a 2×2 matrix:

$$|0\rangle \langle 0| = \begin{pmatrix} 1 \\ 0 \end{pmatrix} \begin{pmatrix} 1 & 0 \end{pmatrix} = \begin{pmatrix} 1 & 0 \\ 0 & 0 \end{pmatrix}$$

The entries of the matrix are telling us the probabilities to find the state in one of the basis states.

```
from qiskit import *
from qiskit.quantum_info import Statevector
from qiskit.visualization import plot_state_city
```

```
#Prepared |Phi+> state:
q = QuantumRegister(2)
c = ClassicalRegister(2)
circ = QuantumCircuit(q,c)
circ.h(0) #superposes first qubit
circ.cx(q[0],q[1]) # First qubit is the control qubit, second qubit is the target qubit
circ.draw('mpl')
```

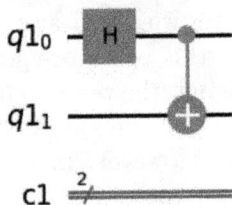

Figure D.5.   Circuit for generating the $|\Phi_+\rangle$ Bell state.

```
state = Statevector(circ)
print(state)
plot_state_city(state)
```

```
Statevector([0.70710678+0.j, 0.       +0.j, 0.       +0.j,
             0.70710678+0.j],
            dims=(2, 2))
```

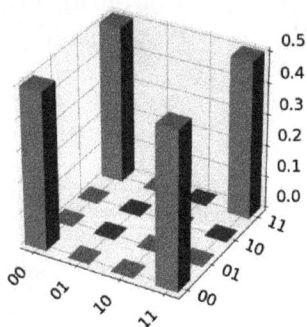

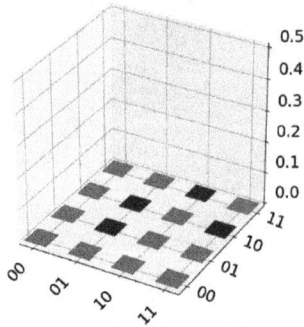

Figure D.6.   Plot state city density matrix plot for the $|\Phi_+\rangle$ Bell state.

Then, let's look at the $|\Phi_-\rangle$ state. The circuit is given in Fig. D.7 and the plot from the "plot state city" function in Fig. D.8.

```
#Prepared |Phi-> state:
q = QuantumRegister(2)
c = ClassicalRegister(2)
circ = QuantumCircuit(q,c)
circ.h(0) #superposes first qubit
circ.cx(q[0],q[1]) # First qubit is the control qubit, second qubit is the target qubit
circ.z(0) # applies a Pauli Z rotation to the first qubit
circ.draw('mpl')
```

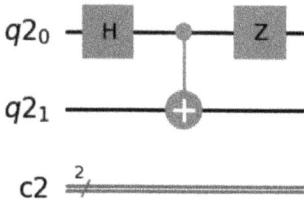

Figure D.7. Circuit for generating the $|\Phi_-\rangle$ Bell state.

```
state = Statevector(circ)
print(state)
plot_state_city(state)
```

```
Statevector([ 0.70710678+0.j, -0.        +0.j, 0.        +0.j,
             -0.70710678+0.j],
            dims=(2, 2))
```

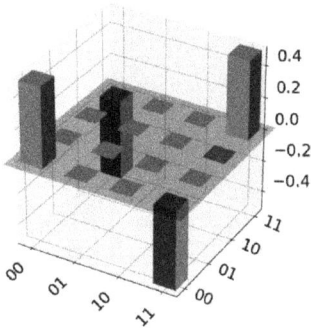

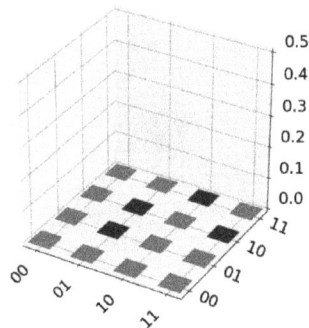

Figure D.8. Plot state city density matrix plot for the $|\Phi_-\rangle$ Bell state.

Then, let's look at the $|\Psi+\rangle$ state. The circuit is given in Fig D.9 and the plot from the "plot state city" function in Fig. D.10.

```
#Prepared |Psi+> state:
q = QuantumRegister(2)
c = ClassicalRegister(2)
circ = QuantumCircuit(q,c)
circ.x(1) #flip second qubit
circ.h(0) #superposes first qubit
circ.cx(q[0],q[1]) # First qubit is the control qubit, second qubit is the target qubit
circ.draw('mpl')
```

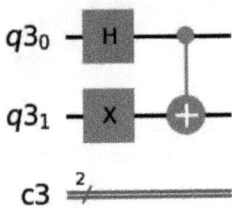

Figure D.9.   Circuit for generating the $|\Psi+\rangle$ Bell state.

```
state = Statevector(circ)
print(state)
plot_state_city(state)
```

```
Statevector([0.      +0.j, 0.70710678+0.j, 0.70710678+0.j,
             0.      +0.j],
            dims=(2, 2))
```

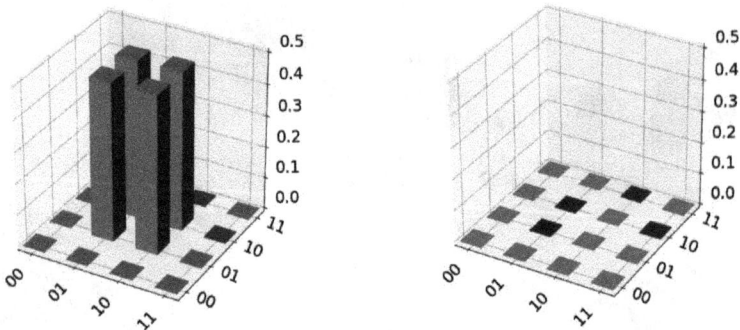

Figure D.10.   Plot state city density matrix plot for the $|\Psi+\rangle$ Bell state.

Finally, let's focus on the $|\Psi-\rangle$ state. The circuit is given in Fig. D.11 and the plot from the "plot state city" function in Fig. D.12.

```
#Prepared |Psi-> state:
q = QuantumRegister(2)
c = ClassicalRegister(2)
circ = QuantumCircuit(q,c)
circ.x(1) #flip second qubit
circ.h(0) #superposes first qubit
circ.cx(q[0],q[1]) # First qubit is the control qubit, second qubit is the target qubit
circ.z(1) #applies a Pauli Z rotation to the second qubit
circ.draw('mpl')
```

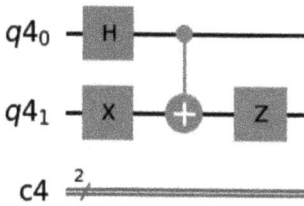

Figure D.11.   Circuit for generating the $|\Psi_-\rangle$ Bell state.

```
state = Statevector(circ)
print(state)
plot_state_city(state)
```

```
Statevector([ 0.      +0.j,  0.70710678+0.j, -0.70710678+0.j,
             -0.      +0.j],
            dims=(2, 2))
```

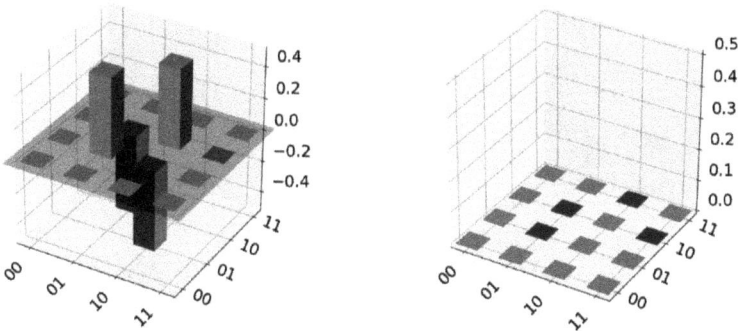

Figure D.12.   Plot state city density matrix plot for the $|\Psi_-\rangle$ Bell state.

2. Recall the formula for concurrence is

$$C = 2\,|c_1 c_4 - c_2 c_3|$$

(a)

$$C = 2\left|\frac{\sqrt{2}}{2}\cdot 0 - \frac{(-1)}{2}\cdot\frac{1}{2}\right| = \frac{1}{2},$$

(b)

$$C = 2\left|\sqrt{\frac{2}{3}}\cdot\sqrt{\frac{1}{3}} - \frac{0}{\sqrt{3}}\cdot\frac{0}{\sqrt{3}}\right| = \frac{2\sqrt{2}}{3} \approx 0.94,$$

(c)

$$C = 2\left|\frac{0}{\sqrt{2}}\cdot\frac{0}{\sqrt{2}} - \frac{1}{\sqrt{2}}\cdot\frac{e^{i\phi}}{\sqrt{2}}\right| = |e^{i\phi}| = 1.^3$$

EPR Paradox

We know that $A(\vec{a},\lambda) = \pm 1$, $A(\vec{b},\lambda) = \pm 1$ and $A(\vec{c},\lambda) = \pm 1$, so $(A(\vec{a},\lambda))^2 = 1$ and so on. Starting from

$$E(\vec{a},\vec{b}) - E(\vec{a},\vec{c}) = -\sum_\lambda \Lambda(\lambda)\left(A(\vec{a},\lambda)A(\vec{b},\lambda) - A(\vec{a},\lambda)A(\vec{c},\lambda)\right)$$

we factor out $A(\vec{a},\lambda)A(\vec{b},\lambda)$ and obtain

$$= -\sum_\lambda \Lambda(\lambda)\left(1 - \frac{A(\vec{c},\lambda)}{A(\vec{b},\lambda)}\right)A(\vec{a},\lambda)A(\vec{b},\lambda))$$

Multiplying by $\frac{(A(\vec{b},\lambda))^2}{(A(\vec{b},\lambda))^2}$, we have

$$= -\sum_\lambda \Lambda(\lambda)\left(1 - \frac{(A(\vec{b},\lambda))^2}{(A(\vec{b},\lambda))^2}\frac{A(\vec{c},\lambda)}{A(\vec{b},\lambda)}\right)A(\vec{a},\lambda)A(\vec{b},\lambda))$$

---

[3]Remember that the magnitude of a phase is 1:
$$|e^{i\phi}| = e^{-i\phi}\cdot e^{i\phi} = e^{i0} = 1.$$

Canceling out $A(\vec{b}, \lambda)$ in denominator and then using $(A(\vec{b}, \lambda))^2 = 1$, we obtain the result we are looking for:

$$= -\sum_{\lambda} \Lambda(\lambda) \left(1 - A(\vec{b}, \lambda)A(\vec{c}, \lambda)\right) A(\vec{a}, \lambda)A(\vec{b}, \lambda))$$

Universal Gate Sets

(a) This gate set is not universal. Let's go through our checklist from Section 6.5.

  1. We cannot create any superposition states with Pauli gates.
  2. CNOT can entangle qubits, but if we cannot create superposition states, this won't be useful!
  3. We have the Pauli Y-gate which can give us complex amplitude states, but again if we cannot have superposition states, this is still not very useful.

(b) This gate set is not universal.

  1. Hadamard can generate superpositions.
  2. CNOT can entangle qubits and since we have superpositions, this is useful.
  3. Hadamard only has real entries, we need another gate that can give a complex phase.

(c) This gate set is universal.

  1. Hadamard can generate superposition.
  2. CCNOT can entangle qubits which can work since we can generate superpositions.
  3. The phase gate allows for states with complex phases.

(d) This gate set is universal.

  1. The rotation gates can generate any superposition states.
  2. CNOT can entangles qubits, which will work since we have superposition states.
  3. The rotation gates take us anywhere on the Bloch sphere, therefore we can access states with complex phase.

Quantum teleportation

1. Let's work step by step to derive $|\psi_2\rangle$. The initial state, $|\psi_0\rangle$, is

$$|\psi_0\rangle = |\psi\rangle \otimes |\Phi\rangle_{AB} = \frac{\alpha}{\sqrt{2}} |0\rangle (|00\rangle + |11\rangle) + \frac{\beta}{\sqrt{2}} |1\rangle (|00\rangle + |11\rangle)$$

We can rearrange to pull out Alice's qubits:

$$= \frac{\alpha}{\sqrt{2}} |00\rangle_A |0\rangle_B + \frac{\alpha}{\sqrt{2}} |01\rangle_A |1\rangle_B + \frac{\beta}{\sqrt{2}} |10\rangle_A |0\rangle_B + \frac{\beta}{\sqrt{2}} |11\rangle_A |1\rangle_B$$

The next step is to apply a CNOT gate between Alice's qubits. Remember that CNOT will flip the target qubit if the control qubit is $|1\rangle$, so $\text{CNOT} |10\rangle_A = |11\rangle$ and $\text{CNOT} |11\rangle_A = |10\rangle$. Thus, we end up with

$$|\psi_1\rangle \, \text{CNOT}_A |\psi_0\rangle = \frac{\alpha}{\sqrt{2}} |00\rangle_A |0\rangle_B + \frac{\alpha}{\sqrt{2}} |01\rangle_A |1\rangle_B + \frac{\beta}{\sqrt{2}} |11\rangle_A |0\rangle_B$$

$$+ \frac{\beta}{\sqrt{2}} |10\rangle_A |1\rangle_B$$

Rearranging again, we get

$$|\psi_1\rangle = \alpha |0\rangle \left(\frac{|00\rangle + |11\rangle}{\sqrt{2}} \right) + \beta |1\rangle \left(\frac{|01\rangle + |10\rangle}{\sqrt{2}} \right)$$

$$|\psi_1\rangle = \alpha |0\rangle |\Phi_+\rangle_{AB} + \beta |1\rangle |\Psi_+\rangle_{AB}$$

Then, Alice applies a Hadamard gate to her first qubit, which gets us to $|\psi_2\rangle$:

$$|\psi_2\rangle = \alpha H |0\rangle |\Phi_+\rangle_{AB} + \beta H |1\rangle |\Psi_+\rangle_{AB}$$

This then gives

$$|\psi_2\rangle = \frac{\alpha}{\sqrt{2}} (|0\rangle + |1\rangle) |\Phi_+\rangle_{AB} + \frac{\beta}{\sqrt{2}} (|0\rangle - |1\rangle) |\Psi_+\rangle_{AB}$$

Expanding this out again and regrouping Alice's qubits together,

$$|\psi_2\rangle = \frac{1}{2} \big[\alpha \, (|00\rangle_A |0\rangle_B + |01\rangle_A |1\rangle_B + |10\rangle_A |0\rangle_B + |11\rangle_A |1\rangle_B)$$

$$+ \beta \, (|00\rangle_A |1\rangle_B + |01\rangle_A |0\rangle_B - |10\rangle_A |1\rangle_B - |11\rangle_A |0\rangle_B) \big]$$

After simplifying, we get

$$|\psi_2\rangle = \frac{1}{2} \big[|00\rangle \, (\alpha |0\rangle + \beta |1\rangle) + |01\rangle \, (\alpha |1\rangle + \beta |0\rangle)$$

$$+ |10\rangle \, (\alpha |0\rangle - \beta |1\rangle) + |11\rangle \, (\alpha |1\rangle - \beta |0\rangle) \big]$$

Table D.1. Possible outcomes of the teleportation protocol.

$m_1 m_2$	Bob's state
00	$\alpha\,|0\rangle + \beta\,|1\rangle$
01	$\alpha\,|1\rangle + \beta\,|0\rangle$
10	$\alpha\,|0\rangle - \beta\,|1\rangle$
11	$\alpha\,|1\rangle - \beta\,|0\rangle$

2. Table D.1 gives the possible measurement outcomes Alice would get based on the state $|\psi_2\rangle$ and the state that Bob receives. So, we can already see that for 00, Bob doesn't need to do any additional rotation.

   (a) For the outcome 01, Bob should do an X-gate in order to flip $|0\rangle$ and $|1\rangle$.

   (b) For the outcome 10, Bob should do a Z-gate in order to correct for the phase of the $|1\rangle$ state.

   (c) For the outcome 11, Bob should do an X- and Z-rotation to get flip to flip $|0\rangle$ and $|1\rangle$ and correct the phase of the $|1\rangle$ state.

Chapter 7

Spin Qubits

1. In this problem, we want to derive the eigenvalues of the general two-level state Hamiltonian given in Eq. (7.2). So, we start with the characteristic equation:

$$\det(H - I E_\pm) = \det \begin{pmatrix} \epsilon - 2E_\pm & \Delta - i\tilde{\Delta} \\ \Delta + i\tilde{\Delta} & -\epsilon - 2E_\pm \end{pmatrix} = 0$$

Computing the determinant and simplifying, we get

$$-(\epsilon - 2E_\pm)(\epsilon + 2E_\pm) - (\Delta^2 + \tilde{\Delta}^2) = 0$$

$$-(\epsilon^2 - 4E_\pm^2) = (\Delta^2 + \tilde{\Delta}^2)$$

$$E_\pm = \pm\frac{1}{2}\sqrt{\epsilon^2 + \Delta^2 + \tilde{\Delta}^2}$$

2. In this case, we can treat the diagonal elements of the Hamiltonian as the z-component, the real part of the off-diagonal elements is the x-component and the imaginary part of the off-diagonal elements is the y-component, as we say in Eq. (7.4), when we expressed the Hamiltonian in terms of the Pauli matrices. We know from relating the Cartesian coordinates, x, y, z, to spherical coordinates that

$$\cos \theta = \frac{z}{r}$$

$$\tan \theta = \frac{y}{x}$$

where $r = \sqrt{x^2 + y^2 + z^2}$. Another way to define θ is to think of the ratio of the projection of the vector on the xy-plane, which is $\sqrt{x^2 + y^2}$ to the z-component, which is $\tan \theta = \frac{\sqrt{x^2+y^2}}{z}$. Substituting in for the components we just identified (i.e. $z \to \epsilon$, $x \to \Delta$ and $y \to \tilde{\Delta}$), we have

$$\tan \theta = \frac{\sqrt{\Delta^2 + \tilde{\Delta}^2}}{\epsilon}$$

$$\tan \phi = \frac{\tilde{\Delta}}{\Delta}$$

Chapter E

Homework Solutions

Homework 1

Complex Numbers

1. Standard form:

$$(1 - i)(6 - 5i) = 1(6 - 5i) - i(6 - 5i) = 6 - 5i - 6i + 5i^2 = 1 - 11i$$

Magnitude: $\sqrt{11^2 + 1^2} = \sqrt{122} \approx 11$

Phase angle: $\cos\theta = 1/\sqrt{122} \longrightarrow \theta = \cos^{-1}\left(\frac{1}{\sqrt{122}}\right) \approx 85°$ or 1.5 rad, clockwise

Rectangular to polar: $1 - 11i = 11\,e^{-i1.5}$

Polar form: First, we find magnitudes of z_1 and z_2 and their relative angles from the real axis:

$$|z_1| = \sqrt{1^2 + 1^2} = \sqrt{2} \approx 1.414$$

$$\theta_1 = \cos^{-1}\left(\frac{1}{\sqrt{2}}\right) = 45° \approx 0.8\ \text{rad, clockwise}$$

$$|z_2| = \sqrt{6^2 + 5^2} = \sqrt{61} \approx 7.81$$

$$\theta_2 = \cos^{-1}\left(\frac{6}{\sqrt{61}}\right) \approx 40° \approx 0.7\ \text{rad, clockwise}$$

Now, we perform the following operation:

$$z_1 \cdot z_2 = \sqrt{2}\, e^{-i0.8} \cdot \sqrt{61}\, e^{-i0.7} = (\sqrt{2} \cdot \sqrt{61})e^{i(-0.8-0.7)} \approx 11\, e^{-i1.5}$$

Converting back to standard form,

$$11\, e^{-i1.5} = 11(\cos(-1.5) + i\sin(-1.5)) \approx 11(0.1 - i) \approx 1 - 11i$$

We can see from the plotted solution in Fig. E.1 that the product of z_1 and z_2 is the result of z_1 being scaled by the magnitude of z_2 ($1.414 \cdot 7.81 = 11$) and rotated by θ_2 which was a clockwise rotation of $40°$. So, our solution ends up at $45° + 40° = 85°$ clockwise. This is what we observe from the figure.

2. Standard form:

$$\frac{7+i}{2+i} = \frac{7+i}{2+i} \cdot \frac{2-i}{2-i} = \frac{14 - 7i + 2i - i^2}{5} = \frac{15 - 5i}{5} = 3 - i$$

Magnitude: $\sqrt{3^2 + 1^2} = \sqrt{10} \approx 3.16$

Phase angle: $\cos\theta = \frac{3}{\sqrt{10}} \longrightarrow \theta = \cos^{-1}\left(\frac{3}{\sqrt{10}}\right) \approx 18°$ or 0.3 rad, clockwise

Rectangular to polar: $3 - i = 3.16\, e^{-0.32i}$

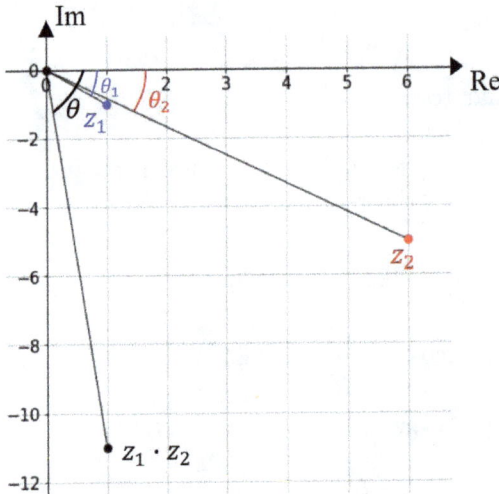

Figure E.1. Plot for Question 1.

Polar form: First, we find magnitudes of p_1 and p_2 and their relative angles from the real axis:

$$|p_1| = \sqrt{7^2 + 1^2} = \sqrt{50} \approx 7.07$$

$$\theta_1 = \cos^{-1}\left(\frac{7}{\sqrt{50}}\right) = 8° \approx 0.14 \text{ rad, counterclockwise}$$

$$|p_2| = \sqrt{2^2 + 1^2} = \sqrt{5} \approx 2.24$$

$$\theta_2 = \cos^{-1}\left(\frac{2}{\sqrt{5}}\right) \approx 26° \approx 0.46 \text{ rad, counterclockwise}$$

Now, we perform the following operation:

$$\frac{p_1}{p_2} = \frac{\sqrt{50}\, e^{i0.14}}{\sqrt{5}\, e^{i0.46}} = \left(\frac{\sqrt{2}}{\sqrt{61}}\right) e^{i(0.14 - 0.46)} \approx 3.16\, e^{-i0.32}$$

Converting back to standard form,

$$11\, e^{-i0.09} = 11(\cos(-0.09) + i\sin(-0.09)) \approx 11(0.996 - i0.09) \approx 11 - i$$

We can see from the plotted solution in Fig. E.2 that the ratio of p_1 and p_2 is the result of z_1 being scaled by the reciprocal of

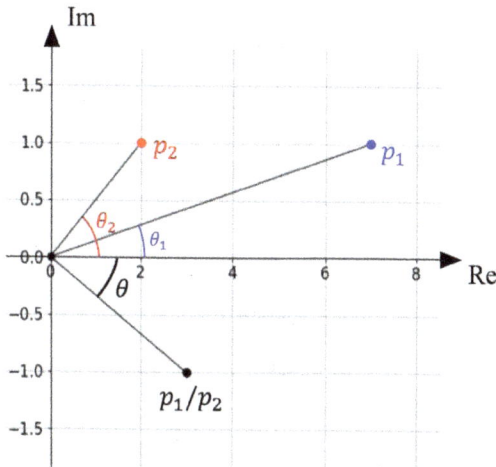

Figure E.2. Plot for Question 2.

the magnitude of p_2 ($\frac{7.07}{2.24}$) and rotated clockwise by θ_2, since we multiplied by the conjugate to simplify the expression. So, our solution ends up at $8° - 26° = -18°$.

3. Standard form:

$$\frac{i(1+i)}{1+3i} = \frac{i+i^2}{1+3i} \cdot \frac{1-3i}{1-3i} = \frac{(i-1)(1-3i)}{1^2+3^2} \frac{i(1-3i)-1(1-3i)}{10}$$

$$\longrightarrow \frac{i-3i^2-1+3i}{10} = \frac{4i+2}{10} = \frac{2i+1}{5} = 0.2 + 0.4i$$

Magnitude: $\sqrt{(0.2)^2 + (0.4)^2} = \sqrt{0.2} \approx 0.45$

Phase angle: $\cos\theta = \frac{0.2}{\sqrt{0.2}} \longrightarrow \theta = \cos^{-1}\left(\frac{0.2}{\sqrt{0.2}}\right) \approx 63°$ or 1.1 rad, counterclockwise

Rectangular to polar: $0.2 + 0.4i = 0.45\, e^{1.11i}$

Polar form: First, we find magnitudes of d_1, d_2, and d_3 and their relative angles from the real axis:

$$|d_1| = 1 \quad \theta_1 = 90° = \frac{\pi}{2} \text{ rad} \approx 1.6 \text{ rad}, \text{counterclockwise}$$

$$|d_2| = \sqrt{1^2 + 1^2} = \sqrt{2} \approx 1.414 \quad \theta_2 = \cos^{-1}\left(\frac{1}{\sqrt{2}}\right)$$

$$= 45° \approx 0.8 \text{ rad}, \text{counterclockwise}$$

$$|d_3| = \sqrt{1^2 + 3^2} = \sqrt{10} \approx 3.16 \quad \theta_3 = \cos^{-1}\left(\frac{1}{\sqrt{10}}\right)$$

$$= 72° \approx 1.3 \text{ rad}, \text{counterclockwise}$$

So, now, we can perform the following operation:

$$\frac{d_1 \cdot d_2}{d_3} = \frac{e^{i1.57} \cdot \sqrt{2}\, e^{i0.785}}{\sqrt{10}e^{i1.25}} = \frac{\sqrt{2}}{\sqrt{10}} \frac{e^{i(1.57+0.785)}}{e^{i1.25}}$$

$$= \frac{\sqrt{2}}{\sqrt{10}} e^{i(2.36-1.25)} \approx 0.45\, e^{i1.11}$$

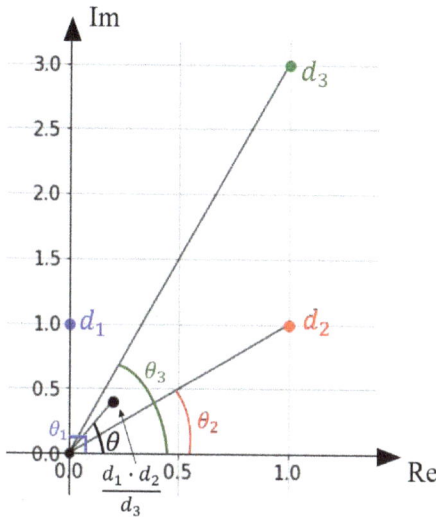

Figure E.3. Plot for Question 3.

Converting back to standard form,

$$0.45 \, e^{i1.11} = 0.45(\cos 1.11 + i\sin 1.11)$$
$$= 0.45(0.44 + i0.9) \approx 0.2 + i0.4$$

As we can see from the plotted solution in Fig. E.3, the final answer results from a mixture of scaling due to division by d_3 and rotations. Let's assume we started with d_2, first we multiplied this number by $d_1 = i$, which caused a 90° rotation counterclockwise, so we had $\theta_2 + 90° = 45° + 90° = 135°$. The magnitude of d_2 was $|d_2| = \sqrt{2}$. Then, we divided by d_3, so the magnitude of d_2 was scaled by $\frac{1}{\sqrt{10}}$ and there was a clockwise rotation of $\theta_3 = 72°$. So, the final answer ended up at an angle $\theta = \theta_2 + 90° - \theta_3 = 63°$, and the magnitude was $\frac{\sqrt{2}}{\sqrt{10}} = 0.45$.

Probability Sample Spaces

1. Since we are replacing placing the ball back in the urn, for the second choice, we may still pick a ball of the same color. Thus are $3 \times 3 = 9$ possible outcomes: BB, BY, BO, YY, YB, YO, OO, OB, OY.

2. When we do not replace the first ball, there are now only $3 \times 2 = 6$ possible outcomes: BY, BO, YB, YO, OB, OY.

Discrete Distributions

1. We are choosing 5 times from the urn, so the possible values that X can take are 0, 1, 2, 3, 4, and 5.
2. X is a binomial random variable since we have a probability of success (choosing a blue ball) to be $\frac{5}{12}$. Therefore, the probabilities associated with each value of the random variable X can be computed as follows:

$$\mathbb{P}(X = k) = \binom{5}{k} \left(\frac{5}{12}\right)^k \left(\frac{7}{12}\right)^{n-k}$$

where k are the values that X can take. Computing each case, we have

$$\mathbb{P}(X = 0) = \binom{5}{0} \left(\frac{7}{12}\right)^5 \approx 0.0675$$

$$\mathbb{P}(X = 1) = \binom{5}{1} \left(\frac{5}{12}\right) \left(\frac{7}{12}\right)^4 \approx 0.241$$

$$\mathbb{P}(X = 2) = \binom{5}{2} \left(\frac{5}{12}\right)^2 \left(\frac{7}{12}\right)^3 \approx 0.345$$

$$\mathbb{P}(X = 3) = \binom{5}{3} \left(\frac{5}{12}\right)^3 \left(\frac{7}{12}\right)^2 \approx 0.246$$

$$\mathbb{P}(X = 4) = \binom{5}{4} \left(\frac{5}{12}\right)^4 \left(\frac{7}{12}\right) \approx 0.0879$$

$$\mathbb{P}(X = 5) = \binom{5}{5} \left(\frac{5}{12}\right)^5 \approx 0.0126$$

3. The plot of the distribution is shown as follows:

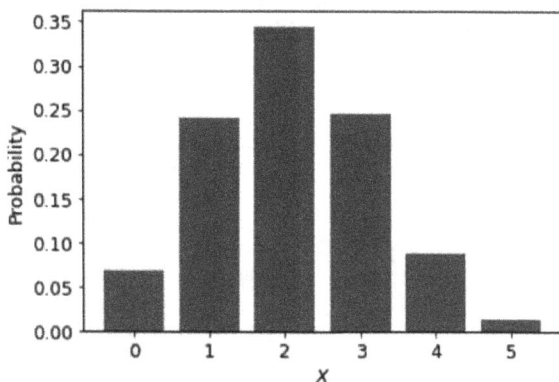

4. Getting at least 3 blue balls means that we can have $X = 3, 4, 5$, so we need to compute $\mathbb{P}(X \geq 3)$, which can alternatively be considered as

$$\mathbb{P}(X \geq 3) = 1 - \mathbb{P}(X \leq 2) = 1 - \mathbb{P}(X = 0) - \mathbb{P}(X = 1)$$
$$- \mathbb{P}(X = 2) \approx 0.3465$$

So, we reframed the question of "at least 3 blue balls" to figuring out the probability of NOT getting at most 2 blue balls. Alternatively, we could have obtained our result from the distribution in the previous part, adding up all the probabilities corresponding to $X \geq 3$.

Calculating Expected Values

1. We are drawing and replacing a ball 2 times from the urn of 5 blue, 4 yellow and 3 orange balls. Now, we can gain \$2 for choosing a yellow ball, lose \$1 for choosing a blue ball, and lose/gain nothing for choosing orange. So, naturally, we can choose a yellow ball 0, 1, or 2 of the times and we can write the sample space: BB, BO, OO, YB, YO, YY. Now, our earnings, E, in \$, can be $-2, -1, 0, 1, 2, 4$.

2. In total, we have 12 balls that we are choosing from, and we want
to find the probabilities of obtaining the choices that we listed in
our sample space. The probabilities are as follows[1]:

$$\mathbb{P}(BB) = \frac{\binom{5}{2}}{\binom{12}{2}} = \frac{5}{33}$$

$$\mathbb{P}(BO) = \frac{\binom{5}{1}\binom{3}{1}}{\binom{12}{2}} = \frac{5}{22}$$

$$\mathbb{P}(OO) = \frac{\binom{3}{2}}{\binom{12}{2}} = \frac{1}{22}$$

$$\mathbb{P}(YB) = \frac{\binom{4}{1}\binom{5}{1}}{\binom{12}{2}} = \frac{10}{33}$$

$$\mathbb{P}(YO) = \frac{\binom{4}{1}\binom{3}{1}}{\binom{12}{2}} = \frac{2}{11}$$

$$\mathbb{P}(YY) = \frac{\binom{4}{2}}{\binom{12}{2}} = \frac{1}{11}$$

If we add up all these probabilities, we get 1, as expected.

---

[1] Note that random variable E is not binomial. This is because we did not specify
what consists of "success," we are just listing all the possible outcomes of drawing
and replacing a ball from the urn twice and what can be earned based on those
outcomes.

3. To calculate the expected earnings, we now have all the ingredients, including the possible values that random variable E can take and the probabilities associated with those values:

$$-2 \cdot \frac{5}{33} - 1 \cdot \frac{5}{22} + 0 \cdot \frac{1}{22} + 1 \cdot \frac{10}{33} + 2 \cdot \frac{2}{11} + 4 \cdot \frac{1}{11} \approx 0.23$$

So, on average, we would earn \$0.23, playing this urn game. Not amazing odds, but it *is* a small win.

Homework 2

1. This question requires using the two different ways we learned for computing the dot product of two Eucledian vectors $\vec{P} = (x_1 \; y_1 \; z_1)$ and $\vec{Q} = (x_2 \; y_2 \; z_2)$:

$$\vec{P} \cdot \vec{Q} = x_1 x_2 + y_1 y_2 + z_1 z_2$$
$$\vec{P} \cdot \vec{Q} = |\vec{P}||\vec{Q}| \cos \theta$$

where θ is the angle between two vectors.

First, we are given the vertices of the triangle, so we must first use the points to form vectors $\vec{AB} = B - A$, $\vec{CB} = B - C$, $\vec{AC} = C - A$. Subtracting the corresponding components of the points, we have that

$$\vec{AB} = (1 \; 5 \; -2)$$
$$\vec{CB} = (2 \; 3 \; -4)$$
$$\vec{AC} = (-1 \; 2 \; 2)$$

The magnitudes for the vectors are $|\vec{AB}| = \sqrt{30}$, $|\vec{CB}| = \sqrt{29}$, $|\vec{AC}| = 3$. Now, we take the following dot products using the first way:

$$\vec{AB} \cdot \vec{AC} = 1 \cdot -1 + 5 \cdot 2 - 2 \cdot 2 = 5$$
$$\vec{AB} \cdot \vec{CB} = 1 \cdot 2 + 5 \cdot 3 - 2 \cdot -4 = 25$$

These must be equal to what we would get by computing the product using the second way:

$$\vec{AB} \cdot \vec{AC} = |\vec{AB}||\vec{AC}| \cos \theta_1 = 5$$

$$\vec{AB} \cdot \vec{CB} = |\vec{AB}||\vec{CB}| \cos \theta_2 = 25$$

Solving for θ_1 and θ_2, we get

$$\theta_1 = \cos^{-1} \left(\frac{5}{3\sqrt{30}} \right) \approx 72.3°$$

$$\theta_2 = \cos^{-1} \left(\frac{25}{\sqrt{30}\sqrt{29}} \right) \approx 32.1°$$

Then, knowing that the total angles in a triangle must add up to $180°$, θ_3, the angle formed between vectors $\vec{AC}$ and $\vec{CB}$ is $75.6°$.

2A. Our two basis states here are $|x\rangle = \begin{pmatrix} 1 \\ 0 \end{pmatrix}$, representing light polarized in the horizontal direction, and $|y\rangle = \begin{pmatrix} 0 \\ 1 \end{pmatrix}$, representing light polarized in the vertical direction. Based on Fig. 2.10, we can see that if the incident light is polarized in the horizontal direction, $|x\rangle$, 50% of the light is reflected and moves in the vertical direction, $|y\rangle$, and gets a $90°$ phase shift, so it's $i|y\rangle$ while the other 50% is transmitted and gets no phase shift, so it remains $|x\rangle$. So, we have some beam-splitter matrix, M_{BS}, that when the initial state is $|x\rangle$, we end up with the following state:

$$M_{BS} |x\rangle = \frac{1}{\sqrt{2}} (|x\rangle + i|y\rangle)$$

Similarly, if the incident light is polarized in the vertical direction, $|y\rangle$, 50% is transmitted without a phase shift and the other 50% is reflected and gets a $90°$ phase shift. So, the beam-splitter matrix gives

$$M_{BS} |y\rangle = \frac{1}{\sqrt{2}} (i|x\rangle + |y\rangle)$$

We can try to guess what the matrix would be by considering that it will be written in the $\{|x\rangle, |y\rangle\}$ basis. So, the first column of the matrix will correspond to the action on $|x\rangle$ and the second will correspond to the action on $|y\rangle$. A more systematic way of finding all the elements is to consider that the 2×2 matrix will have four entries corresponding to $\langle x| M_{\mathrm{BS}} |x\rangle$, $\langle x| M_{\mathrm{BS}} |y\rangle$, $\langle y| M_{\mathrm{BS}} |x\rangle$, and $\langle y| M_{\mathrm{BS}} |y\rangle$. Here, we are keeping in mind that $\langle x|x\rangle = \langle y|y\rangle = 1$, while $\langle x|y\rangle = 0$ since $|x\rangle$ and $|y\rangle$ are orthonormal. Now, we can find the elements of M_{BS} by considering the matrix element:

$$\langle x| M_{\mathrm{BS}} |x\rangle = \langle x| \frac{1}{\sqrt{2}} (|x\rangle + i\,|y\rangle) = \frac{1}{\sqrt{2}} (\langle x|x\rangle + i\,\langle x|y\rangle) = \frac{1}{\sqrt{2}}$$

Doing the same for the other elements, $\langle x| M_{\mathrm{BS}} |y\rangle$, $\langle y| M_{\mathrm{BS}} |x\rangle$, $\langle y| M_{\mathrm{BS}} |y\rangle$, we end up with

$$M_{\mathrm{BS}} = \begin{bmatrix} \frac{1}{\sqrt{2}} & \frac{i}{\sqrt{2}} \\ \frac{i}{\sqrt{2}} & \frac{1}{\sqrt{2}} \end{bmatrix}$$

Note that this matrix is unitary![2] Let's check that this gives us what we expect:

$$\begin{bmatrix} \frac{1}{\sqrt{2}} & \frac{i}{\sqrt{2}} \\ \frac{i}{\sqrt{2}} & \frac{1}{\sqrt{2}} \end{bmatrix} \begin{pmatrix} 1 \\ 0 \end{pmatrix} = \begin{pmatrix} \frac{1}{\sqrt{2}} \\ \frac{i}{\sqrt{2}} \end{pmatrix} = \frac{1}{\sqrt{2}} (|x\rangle + i\,|y\rangle)$$

$$\begin{bmatrix} \frac{1}{\sqrt{2}} & \frac{i}{\sqrt{2}} \\ \frac{i}{\sqrt{2}} & \frac{1}{\sqrt{2}} \end{bmatrix} \begin{pmatrix} 0 \\ 1 \end{pmatrix} = \begin{pmatrix} \frac{i}{\sqrt{2}} \\ \frac{1}{\sqrt{2}} \end{pmatrix} = \frac{1}{\sqrt{2}} (i\,|x\rangle + |y\rangle)$$

Now, let's use the machinery we have learned to find the eigenvalues and eigenvectors of M_{BS}. Let's write down the characteristic polynomial, as we learned in Chapter 2:

$$\det(M_{\mathrm{BS}} - \lambda I) = \begin{vmatrix} \frac{1}{\sqrt{2}} - \lambda & \frac{i}{\sqrt{2}} \\ \frac{i}{\sqrt{2}} & \frac{1}{\sqrt{2}} - \lambda \end{vmatrix} = \left(\frac{1}{\sqrt{2}} - \lambda \right)^2 + \frac{1}{2} = 0$$

Simplifying, we have

$$\frac{1}{\sqrt{2}} - \lambda = \pm\sqrt{-\frac{1}{2}} \longrightarrow \lambda = \frac{1}{\sqrt{2}} \pm i\frac{1}{\sqrt{2}}$$

---

[2]If we check $M_{\mathrm{BS}}^{\dagger} M_{\mathrm{BS}}$, we will see it's equal to the identity matrix!

Now, let's find the eigenvectors:

$$M_{\mathrm{BS}}\,|u\rangle = \lambda\,|u\rangle = \begin{bmatrix} \frac{1}{\sqrt{2}} & \frac{i}{\sqrt{2}} \\ \frac{i}{\sqrt{2}} & \frac{1}{\sqrt{2}} \end{bmatrix} \begin{pmatrix} a \\ b \end{pmatrix} = \frac{1 \pm i}{\sqrt{2}} \begin{pmatrix} a \\ b \end{pmatrix}$$

We obtain two sets of equations for each eigenvalue:

$$\begin{bmatrix} \frac{1}{\sqrt{2}} & \frac{i}{\sqrt{2}} \\ \frac{i}{\sqrt{2}} & \frac{1}{\sqrt{2}} \end{bmatrix} \begin{pmatrix} a \\ b \end{pmatrix} = \frac{1+i}{\sqrt{2}} \begin{pmatrix} a \\ b \end{pmatrix} \rightarrow \begin{matrix} \frac{1}{\sqrt{2}}a + \frac{i}{\sqrt{2}}b = \frac{1+i}{\sqrt{2}}a \\ \frac{i}{\sqrt{2}}a + \frac{1}{\sqrt{2}}b = \frac{1+i}{\sqrt{2}}b \end{matrix}$$

Solving this set of equations, we end up with the eigenvector, $|u_1\rangle = \frac{1}{\sqrt{2}} \begin{pmatrix} 1 \\ 1 \end{pmatrix}$:

$$\begin{bmatrix} \frac{1}{\sqrt{2}} & \frac{i}{\sqrt{2}} \\ \frac{i}{\sqrt{2}} & \frac{1}{\sqrt{2}} \end{bmatrix} \begin{pmatrix} a \\ b \end{pmatrix} = \frac{1-i}{\sqrt{2}} \begin{pmatrix} a \\ b \end{pmatrix} \rightarrow \begin{matrix} \frac{1}{\sqrt{2}}a + \frac{i}{\sqrt{2}}b = \frac{1-i}{\sqrt{2}}a \\ \frac{i}{\sqrt{2}}a + \frac{1}{\sqrt{2}}b = \frac{1-i}{\sqrt{2}}b \end{matrix}$$

Solving this set of equations, we end up with the eigenvector, $|u_2\rangle = \frac{1}{\sqrt{2}} \begin{pmatrix} 1 \\ -1 \end{pmatrix}$. You can check that these vectors satisfy $M_{\mathrm{BS}}\,|u\rangle = \lambda\,|u\rangle$.

2B. Now, we are given a different beam splitter which reflects 60% of incoming photons and transmits 40%. Turning these into fractions, $60\% = \frac{3}{5}$, and $40\% = \frac{2}{5}$. If the state is initially $|x\rangle$, then after going through the beam splitter, the state would become

$$M_{\mathrm{BS}}\,|x\rangle = \sqrt{\frac{2}{5}}\,|x\rangle + i\sqrt{\frac{3}{5}}\,|y\rangle$$

Similarly, if the state is initially $|y\rangle$, then after going through the beam splitter, it becomes

$$M_{\mathrm{BS}}\,|y\rangle = i\sqrt{\frac{3}{5}}\,|x\rangle + \sqrt{\frac{2}{5}}\,|y\rangle$$

Using a similar approach as in the previous part, we obtain the matrix

$$M_{\mathrm{BS}} = \begin{bmatrix} \sqrt{\frac{2}{5}} & i\sqrt{\frac{3}{5}} \\ i\sqrt{\frac{3}{5}} & \sqrt{\frac{2}{5}} \end{bmatrix}$$

Let's check that this matrix works as expected:

$$
\begin{bmatrix} \sqrt{\frac{2}{5}} & i\sqrt{\frac{3}{5}} \\ i\sqrt{\frac{3}{5}} & \sqrt{\frac{2}{5}} \end{bmatrix} \begin{pmatrix} 1 \\ 0 \end{pmatrix} = \begin{pmatrix} \sqrt{\frac{2}{5}} \\ i\sqrt{\frac{3}{5}} \end{pmatrix} = \sqrt{\frac{2}{5}}\,|x\rangle + i\sqrt{\frac{3}{5}}\,|y\rangle
$$

$$
\begin{bmatrix} \sqrt{\frac{2}{5}} & i\sqrt{\frac{3}{5}} \\ i\sqrt{\frac{3}{5}} & \sqrt{\frac{2}{5}} \end{bmatrix} \begin{pmatrix} 0 \\ 1 \end{pmatrix} = \begin{pmatrix} i\sqrt{\frac{3}{5}} \\ \sqrt{\frac{2}{5}} \end{pmatrix} = i\sqrt{\frac{3}{5}}\,|x\rangle + \sqrt{\frac{2}{5}}\,|y\rangle
$$

Homework 3

1. In this question, we use the optical setup for the Mach–Zehnder interferometer in Quantum Flytrap. The first part asks us just to make some observations using the different features. Going to "Waves" mode and using the "Loop" feature, we should observe that with the default setting of the glass, all the light from the laser go to the detector to the right of the last beam splitter in the light path.
2. Now, we right click on the glass and change the phase shift induced by the glass to 0.5 or half-wavelength. We should observe that all the light goes to the detector directly below the last beam splitter in the light path.
3. Now, we need to actually analyze the setup and figure out why we are getting this constructive and destructive interference at the detectors based on the phase shift of the glass. In this part, we go back to the 0 phase shift in the glass. In the right-hand side of the Quantum Flytrap interface, there is information about the wave of light as it goes through its trajectory. The information is given in polar form and consists of an amplitude (normalized to one) and phase, as we learned in Chapter 1. We can select the basis we want to track the state in. In this case, we have linearly polarized light, so we can stay in the default $|\rightarrow\rangle, |\uparrow\rangle$ basis. Starting from the beginning, we have a horizontally polarized light source which we can verify by looking at the right-hand side. We see that the amplitude is 1 and the phase is 0 rad. After the state goes through

the beam splitter, it splits and is thus in a superposition:

$$|\psi\rangle = \frac{1}{\sqrt{2}}(|\rightarrow\rangle + i\,|\uparrow\rangle))^3$$

As the wave travels through the glass and is reflected off the mirror, nothing happens. Finally, the horizontally polarized state is reflected off the mirror and the waves recombine at the beam splitter. As we saw at the first beam splitter, the state that goes through the beam splitter does not pick up a phase shift, but the state reflected off does pick up a $\frac{\pi}{2}$ or 90° phase shift. So, the wave that already has a 90° phase shift will pick up another 90° phase when reflected off, but pick up no phase shift as it moves through the beam splitter. Thus, at the bottom detector there the waves combine, but there is an overall 180° phase difference between them, so they experience perfect destructive interference, while at the detector to the right, both waves have 90° phases which add constructively. This is exactly why we see all the light in the right detector and none at the bottom detector. This confirms what is seen in Fig. 3.11.

4. If we just change the phase of the glass to 0.5, we will observe that most of the analysis will be the same *except* that when the wave moving to the right after the first beam splitter goes through the glass, the glass induces a phase shift of $\pi \approx 3.14$ rad or 180°. So, when the waves meet again at the second beam splitter, the one moving to the right (which had a 90° phase shift because it was reflected off the first beam splitter) picks up an additional 90° phase shift when it is reflected off the beam splitter. The light moving down goes through the beam splitter (picks up no additional phase shift) and has phase 180° because it passed through the glass as it moves to the bottom detector but is reflected and picks up an additional 90° phase shift moving toward the right detector. Now, at the bottom detector, the two waves have the same phase of 180°, so they add constructively, while at the right

---

[3]The $\frac{1}{\sqrt{2}}$ factor comes from the amplitude which is 0.71 since $|1\rangle\sqrt{2} \approx 0.707$. The i factor comes from the fact that $e^{i1.57} = e^{i\frac{\pi}{2}} = \cos\frac{\pi}{2} + i\sin\frac{\pi}{2} = 0 + i$.

detector, one wave has a phase of 270° and the other has a phase of 90°, and the relative difference is 180°, so they destructively interfere. Thus, we see all the light goes to the bottom detector and none goes to the right detector.

5. Finally, if the phase of the glass is $\lambda/4$, the wave that moves to the right after the beam splitter and through the glass acquires a 90° phase shift and when waves meet at the second beam splitter, the waves moving to the right of the beam splitter is reflected and has an overall phase of 180°, but the wave moving down has a phase of 90°, so they do not destructively interfere. The same exact thing happens at the right detector, and thus, we end up measuring 50% of the light in each detector, but the total probability of finding light in one or the other detector adds up to 100%, as expected.

Homework 4

1. First let's import the necessary packages and take advantage of the given helper functions:

```
#Import necessary packages to run code
import numpy as np
from math import pi
from qiskit import *
from qiskit.quantum_info import Statevector
from qiskit.visualization import
plot_bloch_multivector, plot_histogram
from copy import deepcopy

#Helper functions:
def getCounts(circuit, shots):
    """
    Inputs: Quantum Circuit and number of shots
    Returns: dictionary of measurement counts
    """
    simulator = Aer.get_backend('qasm_simulator')
    circ_transpile = transpile(circuit, backend =
simulator)
    result = simulator.run(circ_transpile, shots =
shots).result()
    counts = result.get_counts()
```

```
   return result.get_counts(circuit)

def getExpectationValue(counts, shots):
    """
    Inputs: Dictionary of measurement counts and
number of shots
    Returns: Expectation value
    """
    E = (counts.get('0',0) - counts.get('1',0))/
shots
 return E
```

Based on what we learned in Chapter 4 for doing projective measurements, we will write a function that takes a quantum circuit and converts it into x-, y-, and z-coordinates. We will need to do rotations on the original state to measure along each axis and then take the outcome of the histograms and convert them into coordinates.

Here is what the function should look like:

```
def BlochVectorXYZ(circuit, qr, cr, shots):
    """
    Inputs: QuantumCircuit object, QuantumRegister
object,
            ClassicalRegister object, and number of
shots
    Returns: a vector of x,y,z coordinates of the
QuantumCircuit
    """
    #Cartesian coordinates
    circZ = deepcopy(circuit) #Copy original Quantum
Circuit object
    circZ.measure(qr,cr)
    countsZ = getCounts(circZ, shots)
    z = getExpectationValue(countsZ, shots)

    circY = deepcopy(circuit) #Copy original Quantum
Circuit object
    circY.rx(pi/2,0) #90 degree rotation about X-axis
to swap Z and Y
    circY.measure(qr,cr)
```

```
countsY = getCounts(circY, shots)
y = getExpectationValue(countsY, shots)

circX = deepcopy(circuit) #Copy original Quantum
Circuit object
circX.h(0) #Hadamard gate to swap Z and X
circX.measure(qr,cr)
countsX = getCounts(circX, shots)
x = getExpectationValue(countsX, shots)

return np.array([x , y, z])
```

2. Now, let's test that the function works properly. The first circuit we are told to test consists of a Pauli X-gate and a Hadamard gate. We know that first the X-gate will take us from $|0\rangle$ to $|1\rangle$, and the Hadamard will swap the x- and z-axes, so our Bloch vector will be on the negative side of the x-axis. Using our function, converting to polar coordinates, and comparing with the Statevector output, we get the solution shown in Fig. E.4.

```
#Test case 1
q = QuantumRegister(1) #Initialize Quantum Register
c = ClassicalRegister(1) #Initialize Classical Register for measurement results
circ = QuantumCircuit(q,c) #Create Quantum Circuit
circ.x(0) #Pauli-X gate
circ.h(0) #Hadamard gate
r = BlochVectorXYZ(circ, q, c, 1000)
print(r)

[-1.      0.038 -0.008]
```

```
#Convert to spherical:
phi = np.arctan(r[1] / r[0])
theta = np.arccos(r[2])
state = [np.cos(theta / 2) , np.sin(theta/2)*np.exp(1j*phi)]
state2 = Statevector(circ)
print("From calculation:" , state)
print("From Statevector simulator:", state2)

From calculation: [0.7042726744663604, (0.709417559264796-0.026957867252062246j)]
From Statevector simulator: Statevector([ 0.70710678+0.j, -0.70710678+0.j],
            dims=(2,))
```

Figure E.4. Solution to the first circuit in Problem 2.

```
#Test case 2
q = QuantumRegister(1) #Initialize Quantum Register
c = ClassicalRegister(1) #Initialize Classical Register for measurement results
circ = QuantumCircuit(q,c) #Create Quantum Circuit
circ.x(0) #Pauli-X gate
circ.ry(-pi/4, q) #-45 degree rotation about y-axis
state = Statevector(circ)
plot_bloch_multivector(state)
```

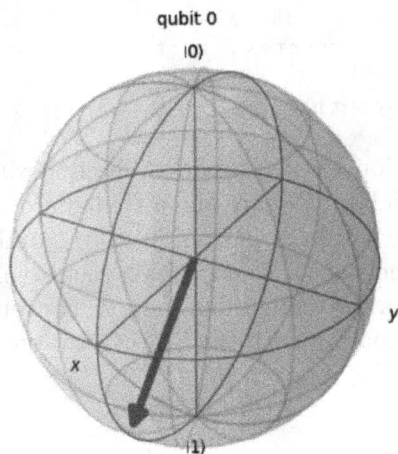

Figure E.5.   Visualization to the second circuit in Problem 2.

We can see that the Statevector output is close to the result we get. The mismatch is due to sampling errors, since we don't have an infinite number of shots.

The second circuit we are told to test consists of a Pauli X-gate and a $-45°$ rotation about the y-axis. Again, the X-gate will first take us from $|0\rangle$ to $|1\rangle$ and the y-rotation of $-45°$ will put us on the XZ-plane. This can be visualized in Fig. E.5.

Using our function, converting to polar coordinates, and comparing with the Statevector output, we get the solution shown in Fig. E.6.

We once again observe that the output our function gives is close to the Statevector output but not exact.

Now, for the last circuit, we have a Pauli X-gate and a rotation of $60°$ about X, which puts us on the YZ-plane. This can be visualized in Fig. E.7.

```
#Test case 2
q = QuantumRegister(1) #Initialize Quantum Register
c = ClassicalRegister(1) #Initialize Classical Register for measurement results
circ = QuantumCircuit(q,c) #Create Quantum Circuit
circ.x(0) #Pauli-X gate
circ.ry(-pi/4, q) #-45 degree rotation about y-axis
r = BlochVectorXYZ(circ, q, c, 1000)
print(r)

[ 0.712  0.028 -0.702]
```

```
#Convert to spherical:
phi = np.arctan(r[1] / r[0])
theta = np.arccos(r[2])
state = [np.cos(theta / 2) , np.sin(theta / 2) * np.exp(1j * phi)]
state2 = Statevector(circ)
print("From calculation:" , state)
print("From Statevector simulator:", state2)

From calculation: [0.3860051813123758, (0.9217841081724197+0.03624993683824123j)]
From Statevector simulator: Statevector([0.38268343+0.j, 0.92387953+0.j],
            dims=(2,))
```

Figure E.6. Solution to the second circuit in Problem 2.

```
#Test case 3
q = QuantumRegister(1) #Initialize Quantum Register
c = ClassicalRegister(1) #Initialize Classical Register for measurement results
circ = QuantumCircuit(q,c) #Create Quantum Circuit
circ.x(0) #Pauli-X gate
circ.rx(pi/3, q) #60 degree rotation about x-axis
state = Statevector(circ)
plot_bloch_multivector(state)
```

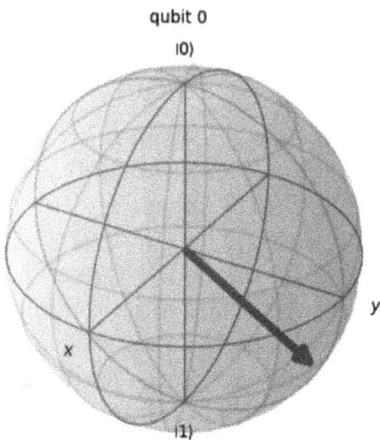

Figure E.7. Visualization of the third circuit in Problem 2.

```
#Test case 3
q = QuantumRegister(1) #Initialize Quantum Register
c = ClassicalRegister(1) #Initialize Classical Register for measurement results
circ = QuantumCircuit(q,c) #Create Quantum Circuit
circ.x(0) #Pauli-X gate
circ.rx(pi/3, q) #60 degree rotation about x-axis
r = BlochVectorXYZ(circ, q, c, 1000)
print(r)
```

```
[ 0.012  0.862 -0.508]
```

```
#Convert to spherical:
phi = np.arctan(r[1] / r[0])
theta = np.arccos(r[2])
state = [np.cos(theta / 2) , np.sin(theta / 2) * np.exp(1j * phi)]
state2 = Statevector(circ)
print("From calculation:" , state)
print("From Statevector simulator:", state2)
```

```
From calculation: [0.4959838707054897, (0.012086973631356013+0.868247605852406j)]
From Statevector simulator: Statevector([0.      -0.5j, 0.8660254+0.j ],
            dims=(2,))
```

Figure E.8. Solution to the third circuit in Problem 2.

Using our function, converting to polar coordinates, and comparing with the Statevector output, we get the solution shown in Fig. E.8.

Now, we see that there is an overall scaling of $i = e^{i\frac{\pi}{2}}$ mismatch between the result we get and the Statevector result. But because this is a global phase factor, this would not affect measurement results!

Homework 5

1. In the first problem, we need to analyze the given table to assess what the privately shared key between Alice and Bob is. Recall the protocol, Alice selects a random bit sequence and randomly chooses a basis to encode the states in. The states are sent via a quantum channel to Bob, who also randomly selects a basis to measure each state in. Finally, Alice and Bob communicate over the classical channel to confirm that Bob received each state and Alice reveals the basis she used to encode the state and Bob also reveals which bases match or not. They immediately discard results measured in bases that do not match. If they knew that there was no one eavesdropping on their communication, this

Table E.1. Homework problem 1 solution.

Alice's random bit sequence	0	0	1	1	0	1	1	0
Alice's basis	Z	Z	Z	X	X	Z	X	Z
Alice's polarization	$|0\rangle$	$|0\rangle$	$|1\rangle$	$|-\rangle$	$|+\rangle$	$|1\rangle$	$|-\rangle$	$|0\rangle$
Bob's basis	Z	X	X	Z	X	Z	X	X
Bob's measurements	$|0\rangle$	$|+\rangle$	$|+\rangle$	$|0\rangle$	$|+\rangle$	$|0\rangle$	$|-\rangle$	$|+\rangle$
Private shared key	0				0	1	1	

would be the end of the procedure and we would end up with Table E.1.

However, they are wary that Eve may eavesdrop, so they actually compare half of the bits that passed the first check. Let's suppose Alice randomly selects and reveals the bits in columns 2 and 7 of the table. They will then see that they get inconsistent results in column 7, so then they know that Eve has actually eavesdropped. Then, they should discard this bit sequence and start over. However, if Alice had chosen a different set of bits, then they may have proceeded without knowing. Obviously, they do not want to compare all the bits because Eve could intercept the classical channel!

(a) So, here, we are assuming that Eve is not eavesdropping, and that there are two bases that Alice is using to encode the information, so there's a 50% chance that Alice and Bob will use the same basis. So, under the assumption of no eavesdropping, their bit string would be 128 bits long.

(b) For Eve to guess the correct bit for one of the bit keys, she would have to measure in the right basis. But she has no idea of the basis until after Bob measures and Alice communicates with Bob. So, whatever basis she chooses, she has a 50% chance to get the right bit string. Remember, if Eve intercepts when Alice sends the bits but measures in the wrong basis, the original information will be lost, so Bob will be able to tell when he compares bases with Alice.

(c) So, for a single bit key, Eve has a 50% chance of guessing correctly, and assuming that each bit key is independent, then the probability of guessing all 20 bit keys is $(0.5)^{20} \approx 10^{-6}$ or

0.0001%. So, having a longer bit string reduces the probability of Eve guessing the key.

2. The following is the code implementing the Elitzur Vaidmann Bomb problem in using Qiskit. Based on the number of rotations we perform, if we have a live bomb, we can prevent the bomb from detonating by continuously forcing the state to be $|0\rangle$.

```
N = 100 #Number of rotations performed
rotation_angle = pi/N #Angle of rotation
shots = 100 #Number of measurements performed

def elitzur_vaidmann_bomb(isBomb):
    """
    Input: boolean object called isBomb.
    1. isBomb = True, means the bomb is live.
    2. isBomb = False, means bomb is a dud

    Output: dictionary of counts from running qasm
    simulator
    """
    #Performing N rotations where each measurement
    will be stored in the Classical Register
    meas = 0
    if isBomb == True: #If we have a Live bomb
        meas = N+1 #Need to keep measuring after each
    rotation to reset the state back to |0>
        else: #Bomb is a dud
            meas = 1 #Only need to measure once

    q = QuantumRegister(1) #Setting up Quantum
    Register with 1 qubit to go through the bomb
    c = ClassicalRegister(meas) #Classical Register
    holds each measurement made after each rotation
    circ = QuantumCircuit(q,c) #Quantum Circuit object

    for i in range(N):
        circ.ry(pi/N,0)
        if isBomb == True: #If the bomb is live
            circ.measure(0,i) #Measurement after each
    small rotation to reset state back to |0>

    circ.measure(0, meas-1) #For a dud, we only
    measure once.
```

```
simulator = Aer.get_backend('qasm_simulator')
circ_transpile = transpile(circ, backend =
simulator)
result = simulator.run(circ_transpile,shots =
shots).result()
counts = result.get_counts()

#Post-processing to make Histogram
predicted_bomb = 0
dud = 0
detonated = 0

if isBomb == True: #Bomb is alive
    predicted_bomb = counts.get('0'* meas) #
predicted bombs but no explosion
    zero_one = '0'*(dud - 1) + '1'
    if zero_one in counts:
        dud = counts.get(zero_one ,0)
    else:
        dud = 0
    detonated = shots - predicted_bomb - dud

else: #Bomb is a dud
    if '0' in counts:
        predicted_bomb = counts.get('0',0)
    dud = counts.get('1')
    detonated = 0

return [predicted_bomb ,dud ,detonated]
```

Now, using the above code, we can make a histogram of the results. If the bomb is not live, we obtain the results shown in Fig. E.9, which is saying that the bomb is not live 100% of the time. If the bomb is live, we obtain the results shown in Fig. E.10. We can see that about 95% of the time the bomb does not detonate even though it's live, and only detonates 5% of the time. If we increased the number of rotations, then this would be further improved.

```
predicted_bomb,dud,detonated = elitzur_vaidmann_bomb(False)
predicted_bomb_prob = predicted_bomb / shots
dud_prob = dud / shots
detonated_prob = detonated / shots
plt.xticks(np.arange(3), ['Predicted Bomb', 'Predicted no Bomb', 'Detonated'])
plt.bar(np.arange(3),[predicted_bomb_prob, dud_prob, detonated_prob])
```

<BarContainer object of 3 artists>

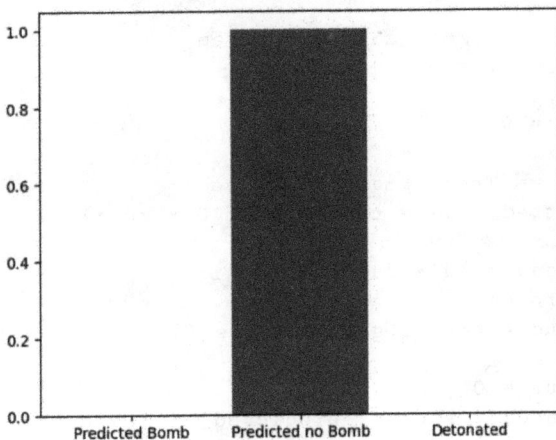

Figure E.9.   Histogram results for a bomb that is not live.

```
predicted_bomb,dud,detonated = elitzur_vaidmann_bomb(True)
predicted_bomb_prob = predicted_bomb / shots
dud_prob = dud / shots
detonated_prob = detonated / shots
plt.xticks(np.arange(3), ['Predicted Bomb', 'Predicted no Bomb', 'Detonated'])
plt.bar(np.arange(3),[predicted_bomb_prob, dud_prob, detonated_prob])
```

<BarContainer object of 3 artists>

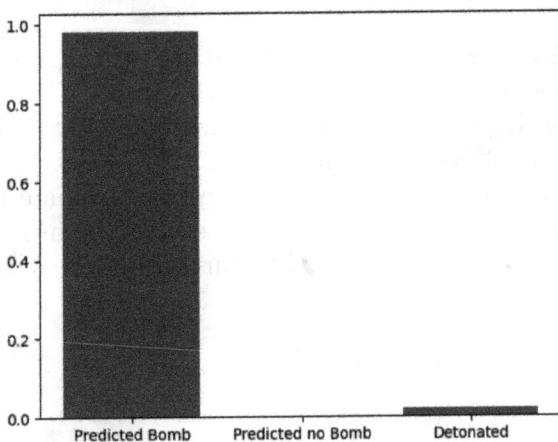

Figure E.10.   Histogram results for a bomb that is live.

Homework 6

1. First, we start by generating the singlet state $|S\rangle$ and the triplet 0 state $|T0\rangle$ in Qiskit which can be done using the following code. First let's do some necessary imports:

```
import numpy as np
from math import *
from qiskit import *
from copy import deepcopy
```

Then, let's prepare the $|T0\rangle$ state. We will need two qubits:

```
#Prepared Triplet 0 state:
q = QuantumRegister(2)
c = ClassicalRegister(2)
circT = QuantumCircuit(q,c)
circT.x(1) #flip second qubit
circT.h(0) #superposes first qubit
circT.cx(q[0],q[1]) #perform CNOT gate
stateT = Statevector(circT)
print(stateT)
```

This will output a vector $\frac{1}{\sqrt{2}}\begin{pmatrix} 0 \\ 1 \\ 1 \\ 0 \end{pmatrix}$. Let's also prepare the $|S\rangle$ state:

```
#Prepared Triplet 0 state:
q = QuantumRegister(2)
c = ClassicalRegister(2)
circS = QuantumCircuit(q,c)
circS.x(1) #flip second qubit
circS.h(0) #superposes first qubit
circS.cx(q[0],q[1]) #perform CNOT gate
circS.z(1) # Pauli-Z gate on second qubit
stateT = Statevector(circT)
print(stateT)
```

This will output a vector $\frac{1}{\sqrt{2}} \begin{pmatrix} 0 \\ 1 \\ -1 \\ 0 \end{pmatrix}$.

Now, we want to figure out a way to rotate the states so that we can determine using a single measurement if we had the singlet or triplet state. If we think about each qubit before doing the entanglement, then we have the second qubit pointing $-\hat{z}$ on the Bloch sphere (because we performed an X-gate), and the first qubit pointing along $\hat{x}$ on the Bloch sphere (because we performed a Hadamard gate). Then, the CNOT gate entangles the qubits. The difference between $|S\rangle$ and $|T0\rangle$ is whether we apply the Z-gate on the second qubit or not. Now, let's imagine we apply a Hadamard gate to each qubit state:

$$HH\,|S\rangle = \frac{1}{\sqrt{2}} \begin{pmatrix} 0 \\ -1 \\ 1 \\ 0 \end{pmatrix}$$

$$HH\,|T0\rangle = \frac{1}{\sqrt{2}} \begin{pmatrix} 1 \\ 0 \\ 0 \\ -1 \end{pmatrix}$$

We can see that the for the $|S\rangle$ state, the Hadamard gate results in a phase rotation of 180° to the original state, while for $|T0\rangle$, it turned it into another entangled Bell state, $|T1\rangle$. For $|S\rangle$, the measurement results are contained to the subspace containing $\{|01\rangle, |10\rangle\}$, while for the $|T0\rangle$ state, the measurements results are contained to the subspace containing $\{|00\rangle, |11\rangle\}$. So, when we do the measurement, we can just check which one of these subspaces we are in to determine if it's the triplet 0 or singlet state.

The following function implements this logic:

```
def singletOrTriplet(circ):
    #Copy original circuit to perform measurement
    meas = deepcopy(circ)
    #Perform Hadamard gate on each qubit
    meas.h(1)
    meas.h(0)
    #Measure each qubit
    meas.measure([0,1],[0,1])

    shots = 1 #single-shot measurement
    simulator = Aer.get_backend('qasm_simulator')
    circ_transpile = transpile(meas, backend =
simulator)
    result = simulator.run(circ_transpile ,shots =
shots).result()
    counts = result.get_counts()
    #extract the keys of the counts dictionary and
take the first key
    state = list(counts)[0]
    # If the two elements of the key are the same
i.e 00 or 11 -> it's Triplet 0 state
    if state[0] == state[1]:
        print('Triplet 0 State')
    #Otherwise, it's the Singlet state
    else:
        print('Singlet State')
```

As an interesting exercise, you can also check whether this function can distinguish $|T2\rangle = \frac{1}{\sqrt{2}}(|00\rangle + |11\rangle)$ from $|S\rangle$. However, for the state $|T1\rangle = \frac{1}{\sqrt{2}}(|00\rangle - |11\rangle)$, two measurements are needed to distinguish it from $|S\rangle$. As a follow-up, if you enjoyed this problem, you can think of a way to distinguish between the 3 triplet states.

2. In this problem, we will analyze the quantum teleportation protocol if we have Eve intercepting. Let's quickly review the protocol: Alice wants to send a qubit state $|\psi\rangle$ to Bob. They already share an entangled state, say $|\Phi\rangle_{AB} = \frac{1}{\sqrt{2}}(|00\rangle + |11\rangle)$, so the initial

state is $|\psi_0\rangle = |\psi\rangle \otimes |\Phi\rangle_{AB}$. Then Alice performs a CNOT operation between her single qubit state $|\psi\rangle$ and her other qubit which is entangled with Bob's. Then, she applies a Hadamard gate on the state she wants to transport and performs a measurement on both qubits, which yields some bit string. She communicates the bit strong with Bob over a classical channel who then knows what additional gate to apply in order to recover the state Alice wanted to send. Let's look back at Section 6.6 to recall the quantum circuit and states that we are dealing with.

(a) For this problem, we are told that Eve is making a measurement after Alice performs the CNOT gate, so Eve is performing a measurement on the state $|1\rangle$:

$$|\psi_1\rangle = \alpha\,|0\rangle \left[\frac{1}{\sqrt{2}}\left(|00\rangle + |11\rangle\right)\right] + \beta\,|1\rangle \left[\frac{1}{\sqrt{2}}\left(|01\rangle + |10\rangle\right)\right]$$

Let's rearrange and simplify by pulling out Alice's qubits because that's where Eve will try to make a measurement[4]:

$$|\psi_1\rangle = \frac{1}{\sqrt{2}}\left[\alpha\,|00\rangle\,|0\rangle + \alpha\,|01\rangle\,|1\rangle + \beta\,|10\rangle\,|1\rangle + \beta\,|11\rangle\,|0\rangle\right]$$

So, if Eve measures Alice's qubits, she has a probability of $\frac{|\alpha|^2}{2}$ to get $|00\rangle$ or $|01\rangle$ and $\frac{|\beta|^2}{2}$ to get $|10\rangle$ or $|11\rangle$. So, by making a measurement, she destroys the state Alice actually wants to send. Eve will in essence teleport $|0\rangle$ or $|1\rangle$ to Bob. Let's say Alice and Bob continue the protocol to the end. If Alice continues the protocol, she would apply a Hadamard gate to the qubit she wants to teleport, but suppose the state has collapsed to $|\psi_1\rangle = |00\rangle_A\,|0\rangle_B$. Then, we would have

$$|\psi_2\rangle = \frac{1}{\sqrt{2}}\left(|0\rangle_A + |1\rangle_A\right)|0\rangle_A\,|0\rangle_B$$

If Alice makes a final measurement to the state, Bob still has $|0\rangle$, so it doesn't make a difference. However, despite her interception, Eve still hasn't learned the state that Alice actually wanted to send.

---

[4]Remember Alice has the superposition state and the first qubit in the Bell states.

(b) If Eve measures the state $|\psi_2\rangle$ before Alice, then she randomly chooses the state that goes to Bob. After Alice measures the state again, it will remain the same, so the correct state will still get teleported to Bob. Eve will know the bit-string outcome from her measurement, so she could know what gate Bob had to do to get the state Alice sent, but she still wouldn't be able to know the state exactly. In this case, Alice and Bob won't be able to tell Eve's interception.

(c) After Alice makes the measurement, the state is teleported to Bob and she cannot recover the teleported state anymore by the no-cloning theorem. If Eve intercepts during the classical communication, she will only know the bit-string outcome of Alice's measurement and thus only know the gate that Bob had to do to get Alice's state, but still would not know what the actual state was!

Bibliography

[1] S. Berryman. Ancient atomism. In E.N. Zalta, (ed.), *The Stanford Encyclopedia of Philosophy*. Metaphysics Research Lab, Stanford University, Winter, 2016.

[2] A.W. Hummel. Science and civilisation in china, volume IV, physics and physical technology, part 1, physics. By J. Needham *et al.*, *The American Historical Review*, 68(2): 463–464, 01 1963. doi: 10.1086/ahr/68.2.463. https://doi.org/10.1086/ahr/68.2.463.

[3] K.S. Krane. *Modern Physics*. Wiley, 3rd edition, 2012.

[4] A. Einstein. On the electrodynamics of moving bodies. *Annalen der Physik*, 17: 891–921, 1905a.

[5] A.J. Greenberg, D.S. Ayres, A.M. Cormack, R.W. Kenney, D.O. Caldwell, V.B. Elings, W.P. Hesse, and R.J. Morrison. Charged-pion lifetime and a limit on a fundamental length. *Physical Review Letters*, 23: 1267–1270, Nov 1969. doi: 10.1103/PhysRevLett.23.1267. https://link.aps.org/doi/10.1103/PhysRevLett.23.1267.

[6] M. Planck. *The Theory of Heat Radiation*. Dover Books on Physics Series. Dover Publications, 1991. https://books.google.com/books?id=eNdD03_92nYC.

[7] A. Einstein. On a heuristic point of view concerning the production and transformation of light. *Annalen der Physik*, 17: 132–148, 1905.

[8] N. Bohr. On the constitution of atoms and molecules. *The London, Edinburgh, and Dublin Philosophical Magazine and Journal of Science*, 26(151): 1–25, 1913. doi: 10.1080/14786441308634955. https://doi.org/10.1080/14786441308634955.

[9] B.R. Wheaton. *De Broglie Wavelength ($\gamma = h/p$)*, Springer Berlin Heidelberg, Berlin, Heidelberg, 2009. pp. 152–154. doi: 10.1007/978-3-540-70626-7_46. https://doi.org/10.1007/978-3-540-70626-7_46.

[10]   C. Davisson and L.H. Germer. The scattering of electrons by a single crystal of nickel. *Nature*, 119(2998):5 58–560, April 1927. doi: 10. 1038/119558a0. https://doi.org/10.1038/119558a0.

[11]   J.D Cresser. Particle spin and the stern-gerlach experiment, *Quantum Physics Lecture Notes*. J.D. Cresser Publisher: Macquarie University, 2009. http://physics.mq.edu.au/~jcresser/Phys301/Chapters/Chapter6.pdf.

[12]   J. Faye. Copenhagen interpretation of quantum mechanics. In E.N. Zalta, (ed.), *The Stanford Encyclopedia of Philosophy*. Metaphysics Research Lab, Stanford University, Winter, 2019.

[13]   A. Einstein, B. Podolsky, and N. Rosen. Can quantum-mechanical description of physical reality be considered complete? *Physical Review*, 47: 777–780, May 1935. doi: 10.1103/PhysRev.47.777. https://link.aps.org/doi/10.1103/PhysRev.47.777.

[14]   J.S. Bell. On the einstein podolsky rosen paradox. *Physics Physique Fizika*, 1: 195–200, Nov 1964. doi: 10.1103/PhysicsPhysiqueFizika.1. 195. https://link.aps.org/doi/10.1103/PhysicsPhysiqueFizika.1.195.

[15]   A. Turing. On computable numbers, with an application to the entscheidungsproblem. *Proceedings of the London Mathematical Society*, 42(1): 230–265, 1936. doi: 10.2307/2268810.

[16]   B.J. Copeland. The Church-Turing Thesis. In E.N. Zalta and U. Nodelman, (eds.), *The Stanford Encyclopedia of Philosophy*. Metaphysics Research Lab, Stanford University, Spring, 2024.

[17]   W. Shockley. The path to the conception of the junction transistor. *IEEE Transactions on Electron Devices*, 31(11): 1523–1546, 1984. doi: 10.1109/T-ED.1984.21749.

[18]   G.E. Moore. Cramming more components onto integrated circuits, reprinted from electronics, Volume 38, Number 8, April 19, 1965, pp. 114 ff. *IEEE Solid-State Circuits Society Newsletter*, 11(3): 33–35, 2006. doi: 10.1109/N-SSC.2006.4785860.

[19]   R.P. Feynman. Simulating physics with computers. *International Journal of Theoretical Physics*, 21(6): 467–488, June 1982. doi: 10.1007/BF02650179. https://doi.org/10.1007/BF02650179.

[20]   D. Deutsch. Quantum computation. *Physics World*, 5(6): 57, Jun 1992. doi: 10.1088/2058-7058/5/6/38. https://dx.doi.org/10.1088/2058-7058/5/6/38.

[21]   P.W. Shor. Polynomial-time algorithms for prime factorization and discrete logarithms on a quantum computer. *SIAM Journal on Computing*, 26(5): 1484–1509, 1997. doi: 10.1137/S0097539795293172. https://doi.org/10.1137/S0097539795293172.

[22]   D.V. Schroeder. *An Introduction to Thermal Physics*. Addison Wesley, 2000. https://books.google.com/books?id=m9_GMAAACAAJ.

[23] C.E. Shannon. A mathematical theory of communication. *The Bell System Technical Journal*, 27(3):379–423, 1948. doi: 10.1002/j.1538-7305.1948.tb01338.x.

[24] S. Axler. *Linear Algebra Done Right*. Undergraduate Texts in Mathematics. Springer New York, 1997. https://books.google.com/books?id=ovIYVIlithQC.

[25] M. Bauer. The stern-gerlach experiment, translation of: "der experimentelle nachweis der richtungsquantelung im magnetfeld", 2023.

[26] D.J. Griffiths and D.F. Schroeter. *Introduction to quantum mechanics*. Cambridge University Press, Cambridge ; New York, NY, third edition, 2018.

[27] C.H. Bennett and G. Brassard. Quantum cryptography: Public key distribution and coin tossing. *Theoretical Computer Science*, 560: 7–11, 2014. doi: https://doi.org/10.1016/j.tcs.2014.05.025. https://www.sciencedirect.com/science/article/pii/S0304397514004241. Theoretical Aspects of Quantum Cryptography — celebrating 30 years of BB84.

[28] A.C. Elitzur and L. Vaidman. Quantum mechanical interaction-free measurements. *Foundations of Physics*, 23(7): 987–997, Jul 1993. doi: 10.1007/bf00736012. http://dx.doi.org/10.1007/BF00736012.

[29] S. Aaronson. *Introduction to Quantum Information Science*, UT Austin, 2018. https://www.scottaaronson.com/qclec/6.pdf. https://www.scottaaronson.com/qclec/11.pdf.

[30] M.A. Nielsen and I.L. Chuang. *Quantum Computation and Quantum Information*. Cambridge University Press, Cambridge, 2010.

[31] C. Couteau. Spontaneous parametric down-conversion. *Contemporary Physics*, 59(3): 291–304, 2018. doi: 10.1080/00107514.2018.1488463. https://doi.org/10.1080/00107514.2018.1488463.

[32] C.H. Bennett, G. Brassard, C. Crépeau, R. Jozsa, A. Peres, and W.K. Wootters. Teleporting an unknown quantum state via dual classical and Einstein-Podolsky-Rosen channels. *Physical Review Letters*, 70: 1895–1899, Mar 1993. doi: 10.1103/PhysRevLett.70.1895. https://link.aps.org/doi/10.1103/PhysRevLett.70.1895.

[33] R. Hanson and D.D. Awschalom. Coherent manipulation of single spins in semiconductors. *Nature*, 453(7198): 1043–1049, June 2008. doi: 10.1038/nature07129. https://doi.org/10.1038/nature07129.

[34] I.I. Rabi, J.R. Zacharias, S. Millman, and P. Kusch. A new method of measuring nuclear magnetic moment. *Physical Review*, 53: 318–318, Feb 1938. doi: 10.1103/PhysRev.53.318. https://link.aps.org/doi/10.1103/PhysRev.53.318.

[35] A. Filler. The history, development and impact of computed imaging in neurological diagnosis and neurosurgery: CT, MRI, and DTI. *Nature Precedings*, July 2009. doi: 10.1038/npre.2009.3267.4. https://doi.org/10.1038/npre.2009.3267.4.

[36] L.M.K. Vandersypen and I.L. Chuang. NMR techniques for quantum control and computation. *Reviews of Modern Physics*, 76: 1037–1069, Jan 2005. doi: 10.1103/RevModPhys.76.1037. https://link.aps.org/doi/10.1103/RevModPhys.76.1037.

[37] J.M.H.L. Sengers. *How Fluids Unmix: Discoveries by the School of Van Der Waals and Kamerlingh Onnes.* History of Science and Scholarship in the Netherlands. Edita, 2002. https://books.google.com/books?id=YwJRAAAAMAAJ.

[38] H. Kamerlingh Onnes. Further experiments with liquid helium. C. On the change of electric resistance of pure metals at very low temperatures etc. IV. The resistance of pure mercury at helium temperatures. *Koninklijke Nederlandse Akademie van Wetenschappen Proceedings Series B Physical Sciences*, 13: 1274–1276, January 1910.

[39] M. Gulka, D. Wirtitsch, V. Ivády, J. Vodnik, J. Hruby, G. Magchiels, E. Bourgeois, A. Gali, M. Trupke, and M. Nesladek. Room-temperature control and electrical readout of individual nitrogen-vacancy nuclear spins. *Nature Communications*, 12(1): 4421, July 2021. doi: 10.1038/s41467-021-24494-x. https://doi.org/10.1038/s41467-021-24494-x.

[40] S.G.J. Philips, M.T. Mądzik, S.V. Amitonov, S.L. de Snoo, M. Russ, N. Kalhor, C. Volk, W.I.L. Lawrie, D. Brousse, L. Tryputen, B.P. Wuetz, A. Sammak, M. Veldhorst, G. Scappucci, and L.M.K. Vandersypen. Universal control of a six-qubit quantum processor in silicon. *Nature*, 609(7929): 919–924, September 2022. doi: 10.1038/s41586-022-05117-x. https://doi.org/10.1038/s41586-022-05117-x.

[41] A. Somoroff, Q. Ficheux, R.A. Mencia, H. Xiong, R. Kuzmin, and V.E. Manucharyan. Millisecond coherence in a superconducting qubit. *Physical Review Letters*, 130: 267001, Jun 2023. doi: 10.1103/PhysRevLett.130.267001. URL https://link.aps.org/doi/10.1103/PhysRevLett.130.267001.

[42] H. Abraham, AduOffei, I.Y. Akhalwaya, G. Aleksandrowicz, T. Alexander, E. Arbel, and A. Asfaw *et al.* Qiskit: An open-source framework for quantum computing, 2019.

[43] E. Schrödinger, Die gegenwartige situation in der quantenmechanik. *Naturwissenschaf Ten* 23: 807–812; 823–828; 844–849, 1935.

Index

www.ingramcontent.com/pod-product-compliance
Lightning Source LLC
Chambersburg PA
CBHW050544190326
41458CB00007B/1904